Humble before the Void

HUMBLE BEFORE THE VOID

A WESTERN ASTRONOMER, HIS JOURNEY EAST, AND A REMARKABLE ENCOUNTER BETWEEN WESTERN SCIENCE AND TIBETAN BUDDHISM

Chris Impey

TEMPLETON PRESS

Templeton Press
300 Conshohocken State Road, Suite 500
West Conshohocken, PA 19428
www.templetonpress.org

Designed and typeset by Gopa and Ted2, Inc.

ISBN-13: 978-1-59947-392-5

Library of Congress Cataloging-in-Publication
Data on file.

Printed in the United States of America

14 15 16 17 18 19 10 9 8 7 6 5 4 3 2 1

Contents

Foreword

My own interest in science began when I was a young boy growing up in Tibet and I came to understand the importance of science as another approach to understanding reality. Not only have I sought to grasp specific scientific ideas, but I have also tried to grasp the wider implications of the advances in human knowledge and technological power brought about through science. Some might wonder, "What is a Buddhist monk doing taking such a deep interest in science? What relation could there be between Buddhism, an ancient Indian philosophical and spiritual tradition, and modern science?"

Although the Buddhist analytical and contemplative tradition and modern science have evolved from different historical, intellectual, and cultural roots, I believe that at heart they share significant things in common, especially in their basic philosophical outlook and methodology. On the philosophical level, Buddhism and modern science both share a deep suspicion of any notion of absolutes. Buddhism and science both prefer to account for the evolution and emergence of life and the universe in terms of interdependence and the principle of causality.

In methodological terms, both traditions take an empirical approach. For example, in the Buddhist investigative tradition, among the three recognized sources of knowledge—experience, reason, and testimony—the evidence of experience takes precedence, with reason coming second and testimony last. In the Buddhist investigation of reality, at least in principle, empirical evidence should triumph over scriptural authority. Even in the case of knowledge derived through reason or inference, it must derive from observed facts of experience. As a result, I have remarked to my Buddhist colleagues that the empirically verified insights of modern astronomy and cosmology compel us now to modify, or in some cases reject, several aspects of the description of the universe found in ancient Buddhist texts.

For more than thirty years I have been engaged in conversations with scientists, led initially by my own personal interest. However, it became increasingly clear that exposure to a scientific worldview would be important and beneficial for my Buddhist colleagues, particularly in the academic centers of Buddhist learning. Consequently, we began to try to introduce formal scientific studies into the monastic colleges.

Humble before the Void is an account of the progress made by the first group of monks to complete the Science for Monks leadership program, an important element in our project to introduce science into the Tibetan Buddhist monastic tradition. It also contains a description of the universe according to the latest scientific research. I trust that this book will also provide readers with a greater awareness of the spirit of curiosity and inquiry that lies at the heart of the Buddhist analytical and investigative tradition, as well as the fruitfulness of maintaining active communication between the Buddhist and scientific communities. I believe these communities complement each other in their attempts to understand reality.

Tenzin Gyatso, the Dalai Lama
April 8, 2013

Humble before the Void

The Edge of the Roof of the World

PEERING INTO DISTANT realms of space is nothing new for astronomers. We wrestle daily with concepts such as infinite space and ancient time. I thought I'd reached the summit of these considerations—until I was invited into another world, where I was forced to deal with time and space in a new and challenging manner.

This was my eight-thousand-mile journey from Arizona to India's border with Tibet, where the Himalayas tower over everything. After a grueling thirty-six hours on the road, nearly half of it in airplanes, my son and I arrived at the "edge of the roof of the world." It was our first time there, and when you arrive, the Himalayas can easily put an individual's life, minuscule by comparison, in perspective.

We cheated the diurnal cycle as we traveled east; the Sun set but then crawled above the horizon as we crossed the Arctic Circle. We rushed headlong into the sunset of the next day and slid into a steamy New Delhi as the lights were going on across the city. Getting to the hotel was a wild fairground ride in a taxi from the 1950s, with weaving traffic, close brushes with auto-rickshaws, and a cacophony of blaring horns. The next morning we completed the journey in a small propeller plane that climbed from the baking plateau and angled toward the crumpled silhouette of the Himalayan range. Cumulus clouds filled the sky, and we were tossed like a child's toy on the descent into the Kangra Valley.

My journey to North India led to an encounter with a remarkable people, the Tibetans in exile. It was also a modern-day encounter between Buddhist philosophy and science. I'd been given the opportunity to teach a select group of three dozen Tibetan monks about modern astronomy, physics, and cosmology. We met under the aegis of the Science for Monks program, an initiative supported by His Holiness, the Dalai Lama. The lives of Western science teachers and Tibetan monks mingled over a three-week workshop of teaching and exchanges, straining all of our notions of time, space, and being in different worlds—East and West, terrestrial and cosmological.

It was a life-changing experience for me and my son Paul, who was about to start his university education. Like most young people he was searching for his vocation and calling in life. Veteran science educators should not put that burden only on students. I found that late in my career, I could also rejuvenate my search for a higher calling and strive for a deeper commitment. In India I was served up a generous helping of questions about life and my priorities. The story in this book is told through my eyes.

At home in Arizona I work at an observatory in the desert, ringed by mountains. I use telescopes with gigantic mirrors and powerful digital cameras, able to snare light from near the beginning of time. During my time near Dharamsala, India, we had only handheld telescopes and the naked eye. But we also had a place at the edge of the rooftop of the world, and a ringside view of the pristine sky that backs the haunting, timeless Himalayas.

This technological and geographical contrast alone could tell much of the story of my experience as a Western scientist immersed in a distant world of Buddhist tradition, which anchors itself in so much that is ancient. The West seeks to understand and control the material world. Buddhism views the material world as, in many ways, illusory.

I wasn't the only person experiencing these stark contrasts. The monks who attended the science classes were also confronted with a world unfamiliar, and in some respects contrary, to their own. The monks and I brought stereotypes—romantic and occasionally derogatory—into our encounter, and one of the most profound experiences we had was to shatter many of them. We also felt like pathfinders. We were engaging in a freewheeling dialogue after centuries of persistent antagonisms between science and religion.

In Arizona I had been curious about Buddhism, and I still consider myself a disinterested, if sympathetic, observer of Eastern religious traditions. I once attended a talk given by the Dalai Lama at our Tucson convention center. It was a huge, impersonal setting, but the small figure on the stage acted as if everyone was in a small, cozy room with him. I was far away in one of the upper-tier seats, and the floor seating was filled with schoolchildren. They were wriggling and whispering and not settling down. With embarrassment I recall my thought at the time: "This is the Dalai Lama. Show him some respect!" I was more preoccupied with the distraction than with what the Tibetan man in maroon robes was saying.

The Dalai Lama is a visionary leader. He and many of his followers have been exiled from Tibet by the Chinese, who have attempted to systematically suppress Buddhism. The Dalai Lama worries that exile is causing Tibetan culture to fade. Although I'm an agnostic, I'm struck by the Dalai Lama's openness to science. In his autobiography he said he would have been an engineer if he hadn't been identified at age two as the fourteenth manifestation of the Bodhisattva of Compassion. He believes Buddhist monks need to be trained for a modern world that's heavily dependent on math and science.

This approach contrasts with the stance of some of the more conservative Christian groups that are active in the United States and in Britain, my native country. The Dalai Lama has met in forums with neuroscientists and philosophers to discuss the nature of consciousness, and the question of doctrine or dogma never arose. Generally speaking, Buddhism has teachings about mysteries and wisdom, but it doesn't assert a theology, especially the formal type known to monotheism.

One could argue that Buddhism has doctrines, yet the central teachings strike me more like a practical form of wisdom. Buddhism believes that humans are temporary vessels of thoughts and emotions, and that the goal of human life is to attain freedom from greed, hate, and ignorance. For Tibetans, this piece of wisdom has proved necessary, since they live in uncertainty and displacement from their homeland.

The same Tibetan Buddhist detachment seems to apply to scientific progress as well. In one famous conversation with the astronomer Carl Sagan, echoed in the foreword to this book, the Dalai Lama said that Buddhism was willing to give up various metaphysical beliefs if they're refuted by clear scientific knowledge. I felt that in the area of science and religion, Buddhism was going to be a welcome partner, so I keenly anticipated my journey to participate in Science for Monks.

When my son and I arrived in India, hopscotching international airports until a small plane took us to the outskirts of Dharamsala, we were met by someone who has been very close to the Dalai Lama's contemporary activity.

Stepping forward to shake our hands was the Venerable Geshe Lhakdor, a monk with walnut-colored skin, high cheekbones, and a broad, lineless face. As with many Tibetans, his age was impossible to guess. He looked forty but the gray flecks in his stubble suggested he was closer to sixty. His face was capable of switching from solemnity to mirth in an instant. To my son and I he had the visage of the seniority expressed in the title *geshe*. Geshe is the most elite designation in the Buddhist academic tradition, the equivalent of a doctorate. It means "spiritual friend." Geshe Lhakdor was also the local host for the Science for Monks program.

At the airport we bundled into his tiny Toyota minivan and headed out on the dusty road toward Dharamsala, the city in northern India that's home to the Tibetan government in exile. Before long we were sitting in Geshe Lhakdor's cramped office in one of the Tibetan government buildings. Outside, a light drizzle had intensified into a pounding rain, a hint of the coming monsoon season. The rain echoed off the corrugated tin roof. The roof was leaking—a by-product of the rapid construction of the buildings, put up after the Dalai Lama had fled Lhasa, Tibet. Such buildings are placeholders for the fabulous Potala Palace in Lhasa, which has a thousand rooms and ten thousand shrines and soars a thousand feet above the valley floor.

Tibetans have been fleeing Tibet ever since the Chinese occupation in 1950. After the failed uprising nine years later, the Dalai Lama escaped to India as well. At that time, his Norbulingka summer palace was hit with eight hundred artillery shells, and two of the three major monasteries in Lhasa were damaged beyond repair. During the difficult period that followed, thousands of monks were executed and many religious artifacts were destroyed. Geshe Lhakdor told us that the rickety Dharamsala buildings were supposed to be temporary. And yet we are still here, he indicated with a gesture.

Geshe Lhakdor served for more than a decade as the Dalai Lama's translator and religious assistant. Although the Dalai Lama is leader of the lineage of Gelug Buddhism, established in Tibet (as the so-called Yellow Hat school) in the sixteenth century, he was also Tibet's head of state for many years. Lhakdor was his go-to guy for travel, public speaking, and arranging accommodations. Now he's director of the Library of Tibetan Works and Archives and thus the chief guardian of Tibetan culture. The library houses eighty thousand manuscripts, many of which are unique. It also holds thousands of photos and hundreds of *thangkas*—finely painted banners of religious motifs. Yet with the ongoing destruction in Tibet, much has been lost forever.

My first day in Dharamsala is still a vivid memory, even though it took place six years ago. Science for Monks has brought Western science educators like me to India twice a year since 2000. We've taught intensive three-week sessions. We've lived in simple and cloistered surroundings. None of us came as Buddhists, as far as I know, and I was as poorly informed as anyone about the subtleties of this ancient religious tradition. Nevertheless, every day I sat cross-legged with monks on mats in a Spartan classroom. We ate simple food and slept and woke with the Sun. We transcended our bodies and projected our minds into the universe. We explored black holes, exoplanets, the interiors of stars, and the early stages of the big bang. We became the geshes of time and space.

In recounting our experiences, this book is part memoir, part journalistic reportage, and part primer on science pedagogy. The scientist in me at first rebelled against writing anything that wasn't as accurate as the account of a scientific experiment. But to tell this very human story, scientific methodology had to take a backseat to the kind of narrative that writers of novels use, sometimes called a *continuous dream*—or, in journalism, *creative nonfiction*. The aim in these pages isn't to depart from facts and reliable eyewitness events. Rather, the goal is to take my six years of traveling to teach in India and condense what I saw and learned into an emblematic three-week experience of Science for Monks (relying mostly on my actual first visit in May 2008).

My chronicle is also the story of an indomitable people who have suffered greatly for their religious beliefs and their culture. They bear their burdens with grace and lightness. The monks in the Science for Monks program have been virtual ambassadors of this kind of character, revealed especially in their commitment to selfless patience and learning.

Over twenty thousand monks are part of the Tibetan diaspora and it takes at least three years to fully train a single small group in science. When the millennial scale of the Science for Monks project was pointed out to the Dalai Lama, he wasn't dismayed. Instead, he gave his signature giggle. Buddhists are experts in taking the long view. The goal is to take monks to a level where they can learn on their own and train other monks in their home monasteries. My time in India was brief, even ephemeral, yet it's a story worth telling, beginning with the memorable first day.

CHAPTER 1

Knowledge and Uncertainty

TRAVELING TWELVE TIME ZONES has turned my body clock upside down. Before teaching the monks I have one day to adjust to India. First impressions are important, and they start flooding my senses. They begin in Dharamsala, where an Indian community coexists with exiled Tibetans. My son Paul and I navigate the small town in a light rain. On top of our jet lag, we're enveloped by the heat and humidity that precede the monsoon season. We try to anchor our bodies and minds in this enchanting yet disorienting environment. At some risk to my Western palate I sample Tibetan tea. It's made with yak butter and salt. The taste is beyond words.

We don't see many people in Dharamsala, to which we will make occasional visits during our three-week stay at the nearby College for Higher Tibetan Studies. Scattered groups of monks walk between the buildings. A few Tibetan men are in Western dress. We learn they are translators for Western visitors. The Tibetan women wear dark, coarse woolen dresses—even in the summer—that they wrap over a blouse. They sport colorful woven aprons if they're married. To keep their hands free, they carry backpacks made from cotton or hemp. With heads lowered, they smile shyly as they pass us.

Local Indian kids have taken advantage of a flat courtyard in Dharamsala to set up an impromptu cricket pitch. For all my existential dread—a new land, a challenging encounter of cultures, a deeply emotional time with my son—my native British heart is warmed. I feel strangely at home when I realize that Dharamsala boasts the highest-altitude cricket stadium in the world.

Our final destination is the College for Higher Tibetan Studies. It's about ten miles down the mountain ridge from Dharamsala, located in rural Himachal Pradesh, a province that extends south to the Kangra Valley floor. The school is a cluster of buildings set into a grove of pine and cedar trees. The cramped site is full of dilapidated buildings. The monks will be living in a campus temple during the three-week Science for Monks program. We'll congregate in separate common areas for eating, and the monks will join me in an unfurnished room for classes from nine to five, six days a week.

As Western visitors, and part of the teaching staff, we'll live in a plain three-story building made of cement and concrete blocks. Down the hall from us is our American host, Bryce Johnson, an environmental engineering PhD out of Berkeley. At age thirty-three Bryce quit a promising postdoctoral position to become organizer of the Science for Monks program. His commitment will sustain us through thick and thin.

Paul and I unpack in our small room. It has a linoleum floor, two cots, and a narrow table. The windows have canvas curtains and ill-fitting screens. We bypass the small wardrobe, deciding it will be easier to live out of our suitcases. The bathroom is bare tile with a shower head coming out of the wall on one side, a small sink, and a toilet with a cracked seat. Toilet paper is thin, unbleached, and pale gray. There's a bucket on the floor for washing clothes.

Outside, the campus is sleepy in summer, with few people beyond those associated with our program. Dogs lie in the shade of most buildings. The school's classrooms and offices fill one of those concrete buildings, and I'll be teaching in a classroom on the second floor, in a space no more than five meters on a side. But the plainness that characterizes the Tibetan enclave also reveals hidden splendors, as Paul and I have already discovered.

Back in Dharamsala we tour the Library of Tibetan Works and Archives. We leave our shoes at the entrance, and red curtains are drawn so we can enter a room suffused with soft, warm light. The room is filled with floor-to-ceiling cabinets. Behind their glass doors are hundreds of scrolls tied with ribbons and encased in wooden planks wrapped in colorful fabric. Some are in Tibetan and others are in Sanskrit. They're the painstaking work of monks over the centuries.

In time I will also see the Tibetan art of making sand mandalas. These symmetrical designs, created on a flat surface with colored sand, are an expression of the Kalachakra, the Tibetan wheel of life. They're exquisitely colored and perfectly formed patterns of sand grains. The monks create them with a small brass funnel that pours sand through a narrow opening. They tap gently on the side to ensure smooth flow, "drawing" on the surface with lines a dozen grains wide. One mandala takes four or five skilled practitioners several days to make. Then, to my amazement, they're swept away in seconds as a powerful reminder of the impermanence of reality.

It takes time for me to understand why such an exquisite undertaking ends intentionally in destruction. In my Western tradition, progress and success are expected. Scientists want to learn new things and build an edifice of knowledge, and we intimately tie our careers and institutions to that quest. All of this seems irreconcilable with the Buddhist precept of evanescence, as illustrated by the sand mandala. Buddhists believe we live in a cosmos with endless cycles of time and existence. Human concerns are insubstantial and impermanent, so the best way to live, according to the Buddha, is to avoid all attachments. Is this Buddhist idea going to have any appeal for me?

At moments like this, I'm reminded of a Western analogy. A century before the Buddha, the Greek poet Archilochus said, "The fox knows many things, but the hedgehog knows one big thing." That, of course, is how to escape the wily fox, making the hedgehog the final victor. Over time I'll come to see that Buddhism has "one big thing" that has allowed it to endure, even prosper, under difficult circumstances, and it is exactly this idea of impermanence and nonattachment. Buddhism sees us as merely temporary vessels of thoughts and emotions, so freedom from worldly attachments is the goal of life. This outlook has an obvious wisdom, and even a psychological payoff. What appeals to me especially is that Buddhism delivers its wisdom without theology or gods. From what I understand, it's also very pragmatic, like science.

On arriving here, I've also recognized that Western science has one big thing: its critical, experimental, and logical method. This may be a kind of attachment, but it seems necessary to overcome our ignorance. I'm waiting to see how this comes across to my new friends, the Buddhist monks. If I have a somewhat romantic view of Buddhism, the same goes for how the monks seem to view science. As I'll soon learn, most of them are new to science. They idealize it as a way to measure everything with perfect accuracy. Science in their eyes is authoritative but also aloof. I hope to teach them that science has a wisdom tradition, too. Like the making of scrolls and mandalas, it can embrace art, play, and mystery.

Our first day has been exhausting. My mental aspirations struggle against an incredible fatigue. Paul and I flop onto our narrow cots. I feel ragged but must try to get some sleep. Tomorrow morning class begins.

Hitting the Mark

Through the drizzle of the coming monsoon, the Sun rises across the campus. It catches mountain peaks and corners of trees and taller buildings. It scatters off the water drops to fill the air with a warm glow. I've arrived at the second-floor classroom to find three dozen monks—at first almost indistinguishable with their bronze complexions, black hair, and maroon robes—ready to learn about science. From the sea of faces, they will emerge as individuals with distinct personalities, and some of them will teach me lessons I couldn't learn anywhere else.

One of these monks is Thupten Tsering. He will become an ally in my efforts to make the science instruction interactive and interesting. I will often call him "Thupten B" during my time in India since Tibetan names are limited in number; there's another Thupten Tsering in the class. You'll get to know Thupten by his appearance: age twenty-nine, he's sturdy like an ox, compact and muscular, with a soft, round face and sleepy eyes. He's also prone to giving bear hugs and lifting others off the ground.

Thupten is from a large monastery in the south, and like many in the region he shows the mild facial scars of tuberculosis. He overcame the disease but it cut short his studies by a year. Now he's two years away from his geshe exam. While not as senior as some monks in the group, he radiates a quiet authority. For me he's a godsend: his English is perfectly articulated, his note-taking superb. He'll be able to help other monks to write journals on what they've learned and add reflections. I can't go wrong with Thupten B. He carries in his bag a hefty and popular Western book on cosmology, Simon Singh's *Big Bang*, but even better, he effortlessly conspires with me to teach the other monks.

For the first day of instruction I begin with something lighthearted, even comical. Thupten helps me orchestrate our first lesson: how scientists deal with uncertainty.

Picture a classroom with a whiteboard, a few windows, and thin pads for sitting. We're all standing, and after we clear an area, I bring Thupten forward, blindfold him, and give him a red marker. He knows what's coming so I spin him around five times. When he stops, he staggers a bit, playing a drunk. It's amusing since Buddhists are teetotalers. I aim him toward the whiteboard. On the board is a black dot at eye level. Thupten walks forward to try to hit the dot. A natural-born actor, he lurches around and careens into the monks standing near a side wall. He stabs the marker at them, but they dodge and duck. He finally hits the wall, but very wide of the mark. More blindfolded monks will try this feat, minus the spinning. But the point is the same: we're coming to grips with uncertainty. Science calls this random error.

The monks take their turns, moving ahead tentatively or with vigor. Some of their marks are close to the target black dot, others are several feet off. Those who miss by a lot receive friendly guffaws. But I tell the monks that this is exactly the way that science—often blindfolded when it comes to measurements of the universe—arrives at as precise an estimate as possible. For example, the monks' overall result shows very little dispersion of the red marks vertically. With fairly good accuracy, they get near the black dot by lining up their hands before moving forward. However, their red marks scatter broadly in the horizontal plane, which they had no fixed way to judge.

I approach the whiteboard and draw a squat oval that includes two-thirds of the marks; this is a classic error "ellipse." Whether wide of the mark or on target, there are no "good" monks and "bad" monks. All the hits are part of a normal error distribution. I imagine the ghost of Carl Friedrich Gauss watching in pleasure; the German wunderkind developed the theory of random errors when he was just a teenager.

Now we do a second version of the experiment. Without being blindfolded, the monks aim magnetic darts at a metal sheet attached to the whiteboard. Everyone has a few practice throws before the one that counts. Some monks come close to hitting the black dot that's the target. Others spray darts across the metal sheet, and a few miss it altogether. The monks quickly realize that they can align themselves horizontally with the black dot but can't help shooting well above or below the target. They have trouble judging the arcing trajectory caused by gravity. I draw an ellipse around the majority of the hit locations, and this time it's elongated vertically. Compared to the blindfolded experiment, the second experiment produces an ellipse that's smaller. Being able to see the target helps the accuracy, but it still produces random error.

When astronomers face the vast universe, I tell the monks, they do something like what we've pulled off in the classroom. Astronomers point their markers and throw their darts, so to speak, and like the monks, they live with the random error associated with any measurement.

During our first day in class we've broken a lot of ice. Our classroom breaks are dedicated to tea time, Tibetan or Western. As the monks walk past me, they flash smiles, a welcoming experience indeed. Their fraternity is obvious. Some walk with arms around each other's shoulders. They're unafraid of physical contact and genuinely seem to like each other. After just one day, the experience of teaching these gentle men is starting to change me. For my break, I go for a brisk walk and begin to take stock.

I'm struck by how completely the monks throw themselves into these experiments. In Arizona I teach hundreds of freshmen, and they arrive at the university with a limited knowledge of science. By contrast, however, they're exceedingly sophisticated in their awareness of social status and the opinion of their peers. To volunteer in class, for example, is generally uncool. The enthusiastic students are in a minority and swimming upstream. Clearly, this isn't the case with the monks. In Arizona, it's hard to find students willing to risk their dignity in a demonstration or experiment. In my humble classroom by the Himalayas, they volunteer with vigor. I'm reminded of the unbridled enthusiasm of fourth-graders back home.

I try not to be too cynical about Western education because it's still the best in the world and I love my job. And yet . . . teachers mostly stay behind a "fourth wall" and try to maintain order and structure. Students mostly accept the peer pressure that favors passivity. In the old Soviet Union, the trope was "They pretend to pay us and we pretend to work." The equivalent in too many American classrooms is "The teacher pretends to teach us, and we pretend to learn." Information is dispersed, memorized, and regurgitated in exams. We hope that it sticks or, even better, that students are "learning how to learn." But the awkward truth is that creativity in the classroom—from either teachers or students—is disruptive and tends to be discouraged.

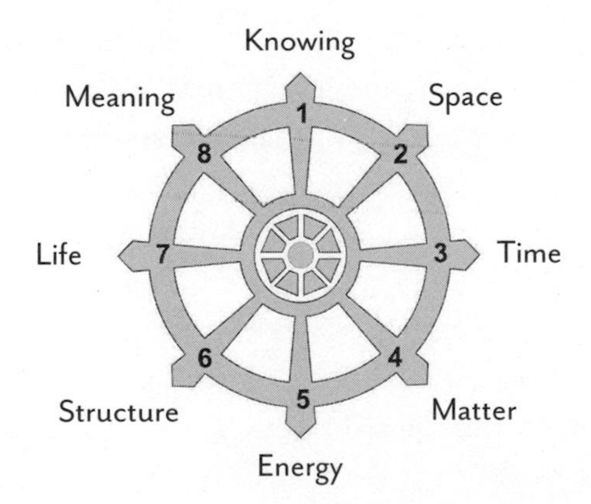

Figure 1.1. The subject matter of the course overlaid on a Buddhist dharma wheel, an allusion to the eightfold path to enlightenment and the eightfold way that led to the quark model of subatomic structure.

The brisk walk is pumping oxygen to my brain, but I don't think my impressions are just euphoria. After one session in this drab classroom in the foothills of the Himalayas, I've had a new kind of teaching experience. I was ready for resistance, but the door was already open. These monks seem happy to accept the fact of human ignorance. If you're not attached to knowledge, learning is simply a gift. They will risk embarrassment to learn. My need to control the class isn't as pressing as usual. The monks are cocreating our lessons. One educational aphorism says, "Learning occurs at the boundary between structure and chaos." I think I'm watching it happen. Before I left Arizona, I was so uncertain about how to present the material that I jettisoned my previous notes and lectures. I started from scratch, a scary move at first, but one that's proving liberating.

By the time I return, the monks are filing back into the classroom. They sit cross-legged on their mats, and we continue our discussion of error and uncertainty. My translator today, and for most of the days to come, is Tenzin Sonam. Tenzin has a degree in engineering and is on staff at the Library of Tibetan Works and Archives. He's not a monk. You'll know him, too, by his appearance. In his late twenties, he wears T-shirts and chinos, and he has the high forehead of a thinker and the milky coffee complexion of a Tibetan. He's a superb translator between English and Tibetan, a process that still amazes me.

I tell Tenzin we're going to cement our earlier lesson. Each monk receives a wooden ruler in millimeters and I ask them to measure the width of the whiteboard. It's about eight meters across, so they have to place the rulers carefully end to end twenty-five times. As they finish I collect the answers and make a histogram, which is a kind of chart of results, ideally looking like a bell curve in the end. Our result is slightly ragged but it's also pleasingly symmetric, with a mean of about 785 centimeters and a scatter of about four centimeters.

"Who got the answer right?" I ask.

Half the monks put their hands up, and the other half guffaw and hoot with mock derision. I notice they like to joke with each other about such conceits as, "Who got the right answer?" "Who throws best?" and by implication, "Who is the best monk?" So I play on this, because it's a good lesson in science as well. With random error there's no single "best" result, only averaging of intrinsically uncertain measurements.

"Who's the best monk?" I ask.

They look sidelong at each other and remain silent. With the markers and darts, we've learned that the good and bad throws are really just the same in a random scheme. I also suspect they prefer to avoid the brazen ego associated with claiming to be "the best."

The averaging of the values in the histogram, however, gives an estimate of the true value. So I ask, "What's the best answer?" Now there's agreement. They all point to the numbers at the peak of the histogram, the high point of the bell curve. This is how we make the best measurements in science, I tell the group.

In astronomy, I explain, it makes no sense to ask where exactly in the sky a star is. If I gave each of you an identical telescope and asked you to measure the two angles that define the position of any star on the sky, which are the equivalent of X and Y on a plane, you'd each get different answers. The "best" position would be given by the average of each set of angles. The tolerance or uncertainty in the position would be given by the scatter or spread in each set of angles. It's possible that no individual measurement with a telescope will match the best position. A murmur of approval ripples around the room. They like this egalitarian principle, since Buddhism teaches that people must contribute to the common good.

A question comes from a tall and slender monk over to my side. "What if you use a better telescope?" he asks. His name is Gelek Gyaltsen, and as I will learn later, he's from a remote village in eastern Tibet. As the days pass, I will hear stories of many of the monks, including Gelek. For now, his long arms, delicate hands, and dark eyes make a formidable impression. His stare seems slightly intimidating. He's no stranger to Western science pundits like me because he's already studied in the United States. He later tells me, "I noticed a huge difference between students there and monks here in the monastery. People there tend to have sudden emotional changes, often for trivial matters; they easily get too excited or too disappointed. Monks don't react that way; they neither get too elated nor too sad." They inhabit the middle of the bell curve.

"A better telescope?" I reply. You'd get a tighter cluster of data points. The best position might shift but there would still be uncertainty in the measurement. Even if you could get your hands on the best telescope money could buy, the measurement of a position would be limited by the blurring of Earth's atmosphere. And if you spent even more money to put a telescope in orbit, the image would be sharp, but the light path would be slightly bent by gravity along the sight line. You'd still be slightly unsure of the star's position.

Gelek Gyaltsen nods. For all the monks, the impossibility of a perfect measurement in astronomy is sinking in.

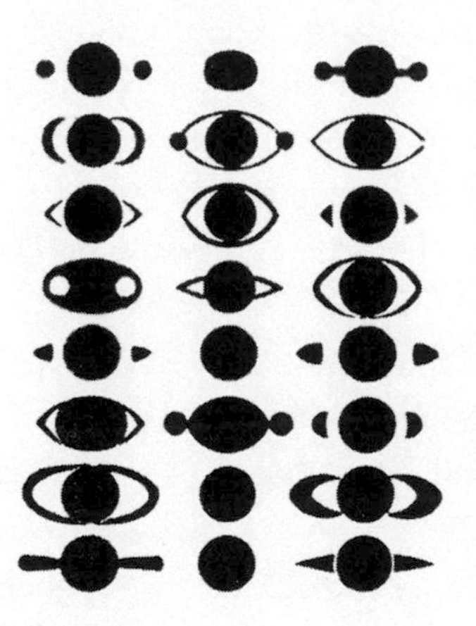

Here are observations of Saturn made in the first fifty years after the telescope's invention, from 1610 to 1660.

Which one shows the way Saturn "really looks"?

Science progresses at the limit of observation. Scientists always want more and better data.

Figure 1.2. Observations always have limitations and at the limit of observation, interpretation is uncertain. Plato's supposition of an underlying objective reality is not meaningful in modern science.

What this means in practice, I continue, is that a single measurement with a telescopic instrument is only "good" at best. Better would be multiple measurements with that instrument. However, the best is this: independent measurements by various investigators using different instruments. You can make a careful measurement yourself, but until you see the measurement repeated independently, you don't really know how accurate it is.

Twenty minutes later, with the taste of a sweet-tea-and-biscuits break on our lips, we return, and I take the class through one more exercise. I hope they find it a challenge and will be able to follow my story line. It's a kind of play acting—what science calls a "thought experiment." I announce that I'm an alien from outer space who has come to Earth. My mission is to report back to my planet what earthlings are like. During my data collection I list human features: two eyes, two ears, two legs, and two arms. This strongly suggests the earthlings are symmetrical. They have two of almost everything. But then I notice that they tend to make scratchy marks—human writing—with just one arm. This isn't symmetrical behavior. My faithful translator, Tenzin, tries to get this across.

One by one I ask them, "What arm do you use to make scratchy marks?"

Through the first and second rows, everyone uses their right hand to write. So, I ask them, can I report to my home planet that the symmetry of earthlings is an illusion, since they only use the right arm to make their scratchy marks?

Someone in the back throws a wrench in the works. He's Dawa Dorjee, and he waves his left hand in challenge to my statistical report. "This is the hand I write with," he says. Two more left hands go up in the crowd. I return to my role as the alien investigator. I must have done my statistics badly, I say. I concluded that earthlings only use the right side because I asked only twelve monks. Three left-handed monks were left out of my data. The lesson? Sample size and selection are core issues in any kind of data gathering. If a phenomenon is rare, a lot of data is needed to know how often it occurs.

The ruse works pretty well. We're now discussing how much data is enough if you want to look for small or subtle effects. Then we segue into a freewheeling discussion of why humans are dominantly right-handed. Several monks say they were naturally left-handed but were "persuaded" to use the right hands by parents or teachers. Indeed, one of them is ambidextrous. He opens his journal, borrows a pen, and writes on both pages at once, each hand a perfect mirror of the other. I'm exhausted but I marvel at this sight. I doubt this monk knows that Leonardo da Vinci was famous for writing this way. We break for the day.

The Importance of Evidence

- There is no science without evidence.
- All assertions must be supported by data.
- Every claim in science is subject to verification.

Science is data-driven, so progress is made by:

1. Gathering more data	**Good!**
2. Repeating the experiment	**Better!**
3. Someone else repeating the experiment	**Best!**

Figure 1.3. Science is fundamentally empirical, resting on ceaseless observation and experimentation of the natural world. Limitations of the evidence can lead to areas of disagreement, but ultimately to progress.

That night my role-playing as an alien comes back to haunt me. The jet lag and cultural dislocation have made my body feel like it truly has landed on another planet. A falling brick of fatigue hits me at various points in the day and then resurfaces as insomnia at night. On the first night, back at our room, Paul has fallen asleep at dinner time. With the capacity of a teenager, he won't stir the entire night. As for me, the first glimmer of sunrise trickles through the window before I roll into a deep sleep. The only consolation is that I don't have to teach again until the afternoon. When I wake the Sun is high and pouring through a gap in the canvas window covering. Groggy, I splash my face with water, make myself presentable, and head downstairs just in time to catch lunch.

This is indeed my planet, stories of aliens notwithstanding. At the lunch table I see my four Western colleagues in Science for Monks:

Bryce Johnson, our organizer; Gail Burd, who's teaching the biology section; and Mark St. John and Richard Sterling, our two evaluators and education analysts. They're feasting on noodles mixed with vegetables and popping into their mouths a chief staple of Tibetans: momos. Momos are dumplings filled with lamb or vegetables and dipped in soy sauce. In a Tibetan home, the meat would be yak. Tibetans also like beef, but since cows are sacred through most of India, they settle for mutton.

As I join them, Gail is to my left. She has the look of a new initiate. Like me, it's the glazed look of travel. She pushes the noodles around her plate without much interest. I feel a vested interest in Gail having a good experience; I helped recruit her, and she's the associate dean of science at my university. In contrast, Mark and Richard look chipper. They've been here for a week. Mark has red hair, a droopy mustache, and a soft Californian accent. He'll be evaluating the program. Richard is somewhat older, with short, gray hair, a crisp English accent, and a military bearing. He's going to oversee the monks keeping journals and writing about their learning experiences and their perspectives on Buddhism and science.

Bryce is our natural interlocutor. He has the big picture, and from extensive experience he knows how to survive in India better than anyone. As he tucks into momos, he dispenses useful tips: where to hang wet laundry to dry, interesting local temples, auto-rickshaw rates, unwritten rules of haggling in street markets, and how to find the best coffee in Dharamsala. With a pale complexion set off by bushy eyebrows and a shock of thick black hair, Bryce has a refreshingly casual attitude to the inefficiencies and routine frustrations of trying to do business in rural Himachal Pradesh. He's never flustered.

One thing that even Bryce can't do for us is improve the infrastructure. In this rural province, cows may be sacred but electricity is not. It goes out daily, often for hours at a time. When it's working, the current surges so much that we fear for our gadgets. Plugs spark and crackle every time they connect to a wall outlet. There's Internet, but it's slow and fitful, and the connection often times out before a web page can load. I will suffer intense Internet withdrawal for the first few days.

Practical as usual, Bryce has given us flashlights and candles, which we use daily. But there's little he can do about hot water. There are solar units on the roof, but by the second day something has gone wrong. We take cold showers the rest of the trip. We flinch on chilly mornings, but on blistering afternoons it can be a pleasure.

As we adjust to low technology, the beautiful scenery of India is a welcome distraction. When Paul and I look out our bedroom window, we see a lush landscape of vines and mangroves, rhododendrons and acacias. Below us spreads a mosaic of rice paddies etched onto the fringes of the Kangra Valley. Above us are pine groves and neatly pruned tea plantations clinging to the undulating contours of the hillside that rises up to Dharamsala.

With the flora of the countryside also comes the fauna. Besides monkeys on the edge of town, and dogs everywhere, the fauna is mostly insects. They come by land, sea, and air—through cracks in the walls, up the bathroom drain, and via gaps under the door and at the edges of the window screens. Mosquito coils and repellant do little to stop them, and they are legion: centipedes, millipedes, dragonflies, wasps, ladybugs, cockroaches, tiny red fire ants that pack a blistering sting, bugs that startle when they fly past with a loud whir, and three-inch long beetles with iridescent blue-green carapaces.

To top it all off, an early monsoon has arrived in full force. Every day after class the sky crackles and rumbles before delivering a deluge. Every morning we wake to the din of rain thrumming off the tin roof. Water, water everywhere, but drinking it is hazardous. Paul and I stick to bottled water, and we're careful to keep our mouths closed while showering. Nothing dries, nothing heals.

The Art of Estimation

The morning after our second night I return to the classroom with great expectations—even if our topic today is mathematics, a hazardous bridge to cross even at American universities. Buddhist monastic tradition involves extensive instruction in theology, philosophy, rhetoric, and history. But math is not a central part of the curriculum.

As well as I can judge, most of the monks have only reached the level of a seventh- or eighth-grade student back home. They exhibit an unusual combination of technical naivety and conceptual sophistication.

Scientists use mathematics as their central tool. They're not afraid of algebra or calculus, and most aren't even intimidated by second-order differential equations (to be totally honest, the latter can induce slightly sweaty palms even in seasoned professionals). But showy math isn't always required in science. The simpler skill of estimation is equally important. An artist might be able to painstakingly render a human face with layers of oil paint but the ability to capture a likeness with a few strokes of pencil or charcoal is also prized.

Precision isn't always possible in astronomy, I tell the monks, picking up where we left off. Astronomical objects are remote and unreachable. The best-determined numbers are known with an accuracy of 10 percent while others are uncertain by a factor of two. In other words, the true answer might be as much as two times higher or as little as two times lower than the estimate. The art of estimation allows a scientist to learn something roughly but reliably, when learning it precisely is impossible. Among the sciences, astronomy works most routinely with the problem of large numbers. But biologists are uncertain of their estimate of 10 million plant and animal species, and neuroscientists have to estimate the number of neural connections in the brain, which may be as large as 1,000 trillion.

Poets deal with large numbers by speaking of "all the stars" in the sky or "every grain of sand" on the Earth. I tell the monks that astronomy also inhabits the realm of incredibly large numbers, and for many people this simply becomes too abstract to comprehend. My goal is to give them a visceral sense of massive numbers and how astronomers make these estimates, since nobody can actually count the stars in the sky or grains of sand on the beaches.

I ask the monks, "What's the biggest number you know not just in your head, but in your real experience?"

They ponder this for a time, and the responses range from four thousand monks at the largest monastery near Bangalore, where they occasionally gather in a single room, to 6 million Tibetans, which is a bit more abstract but still conceivable. We all have our first experiences with large numbers. Mine was a child when I collected pennies and turned them into dollars. Over the years, my comprehension grew with such experiences as watching a hundred thousand rock concert fans wave hand lights at Wembley Stadium in London, or walking beside a wheat field with my grandfather in the Scottish Borders. With his help I estimated that the field contained 100 million ears of wheat—a truly impressive number.

To take the monks down this path, I use two items in countable reality that they know very well: books and sand. They revere books for their ancient wisdom, and sand is the medium by which they produce mandala paintings, only to sweep them away. First, I ask them to imagine all the books ever written. How many pages would be in all these books and how high would the pile be? We do simple scaling and try to base it on their real experience (not an Internet search, which wouldn't be possible anyway given its flakiness here). They begin estimating the number of books at the Library of Tibetan Works and Archives, but the issue is settled by Tenzin, who works there. There are eight thousand books and three thousand scrolls.

Tenzin elaborates further. If Tibetan culture had not been systematically destroyed, the number of books would be more like a hundred thousand. So we use that number to make an estimate of the total number of books in the world, presuming that Tibetan book generation is typical of other cultures. If Tibetans have created one hundred thousand books and they're 0.1 percent of the world's population, then the whole world has produced roughly 100 million books. The fact that some works have been translated into multiple languages would reduce this number slightly.

I suggest that books average two hundred pages, and the monks note that their notebooks have one hundred pages and are a centimeter thick. Two centimeters per book, times 100 million books, gives two thousand kilometers. Therefore, a shelf holding all books ever written would stretch across India from north to south. Reading them all at the rate of a book a day would take 250,000 years. I hear murmurs of satisfaction at this number; the monks are dedicated readers. These speculations about books have no practical value but they show the monks how to go from numbers they "know in their gut" to numbers that are meaningful but almost imponderable.

Science Limitations

Uncertainty, imprecision, and error arise three different ways:

Conceptual	Making a false premise, confusing correlation with causation, inferring a pattern where none is present.
Macroscopic	There is no such thing as perfect data. Every data set is limited and every instrument has limitations.
Microscopic	Heisenberg's uncertainty principle sets a fundamental limit to precision for measurement of particle position and velocity, or energy and time.

Figure 1.4. In science, and in astronomy particularly, estimation is a skill by which scientists develop their intuition. There is a fundamental limit to our knowledge of the microscopic world of the atom.

Sand is definitely something they know in their gut. To make a mandala, monks almost literally put the colored grains down one by one, a project best characterized as a labor of love. Today's discussion of humble silica will be our first, but I'll use sand frequently in the days ahead. It will be a microcosm and a metaphor for large numbers, and later on, it will stand for distant worlds in space. We begin, though, with a mere bucket of sand, which I uncover and carry to the center of the classroom. Our goal is to estimate how many grains are in the bucket. The stepping-stone will be their gut-level familiarity with the amount of sand in a mandala. I begin by asking the monks to calculate how many grains are in a cubic centimeter, about the size of a sugar cube. They do this with a ruler and I'm happy to see that it becomes a project in peer learning.

The monks are in six groups based on their home monasteries. I plan on routinely breaking the class into smaller groups to do experiments and hands-on activities. I consider forming the groups randomly, but monastery groups make sense since they know each other already, and I want to forge working relationships that they can benefit from after the workshop is over.

They don't have to fill a tiny cube with grains one at a time; I tell them how to take a shortcut. In each group, the monk with the best eyes and the steadiest hand lines up sand grains on the floor, just touching, until they span a centimeter. This number cubed approximates the number of grains in one cubic centimeter. My students back home would struggle with this activity, not because it's mathematically difficult, but because it's painstaking and tedious. The red-robed men perform the task without a hint of self-consciousness, and with care and delicacy, like a devotion.

I caution them about errors in the method. They're lining up grains edge to edge, but this is artificial. In any real volume of sand, the grains don't sit in neat rows and layers; they settle and scrunch together. Furthermore, the monks have understandably avoided the tiniest grains in the bucket. Both these tendencies will make their final estimate of the number of grains in a cubic centimeter too low.

My mention of variations in sizes and configurations of sand sets off a spirited debate, a debate that probably only Tibetan monks would enter. It escapes me for a moment until Tenzin translates. The monks know that mandala sand is by no means ordinary. It's high-grade quartz sand, made by grinding larger rocks until the fragments reach a nearly uniform size. The sheen and sparkle of quartz is part of what gives a sand mandala its allure. All I could get hold of was builder's sand, which is crude. Rubbed between the fingers the smallest particles are little more than dirt. I watch as the monks heatedly argue over sand. They engage in a little jostling and shoving. Back home, three dozen men would only get this worked up over sports.

As Tenzin translates I begin to catch on. They're not arguing over who knows mandala sand best but how even the coarse sand I brought averages out in its small and large pieces to be about the same average size as a crystal of high-quality mandala quartz. Once I hear this I feel a surge of pride. They're displaying perfect scientific form, weighing systematic errors like hard-nosed professionals.

They follow geometric logic to calculate the number of grains in the bucket. The sand mandala, being a circle, is like a slice of the cylindrical bucket, and they figure that the slice is ten grains of sand tall. Each group now has a height for the bucket and the mandala, so they begin to calculate. They emerge from their groups, average their six different conclusions, and arrive at a consensus: 80 million grains in a mandala, and therefore 700 million in the bucket. They now have a gut-level feeling for numbers like 100 million and even 1 billion.

This kind of calculating catches on as the monks try to make it personal and tangible. In fact, that's the assignment they're to bring back for the next morning's class. First up is a monk named Sherab Tenzin, and for his homework he estimates that he's taken 18 million breaths in his life. I like his example. It's concrete yet evanescent—those breaths have merged seamlessly into the air, leaving no trace of their existence.

Then Gelek Gyaltsen announces that he eats 7 million grains of rice a year. When I ask him about the population of India, he makes this calculation on the whiteboard: sixty thousand grains in a cup, three meals a day, and a billion Indians. As he works, I think that my university students in this situation would turn to Google or a calculator. The monks are low tech but they also have the patience for a hands-on approach. Still, they struggle with numbers when there are too many variables. But Gelek, with a little help from the sidelines, finally gets there: Indians eat 7,000 trillion, or 10^{16}, grains of rice a year. He looks around at the class and flashes a megawatt smile. It's a prodigious number.

The monks like working with large numbers at a gut level. Soon they will realize that even best scientists rely on a similar intuition of experience in approaching higher theoretical realms of physics. I have a colleague who's a gravity theorist; I once asked him, "How many people understand general relativity?" General relativity is Einstein's theory

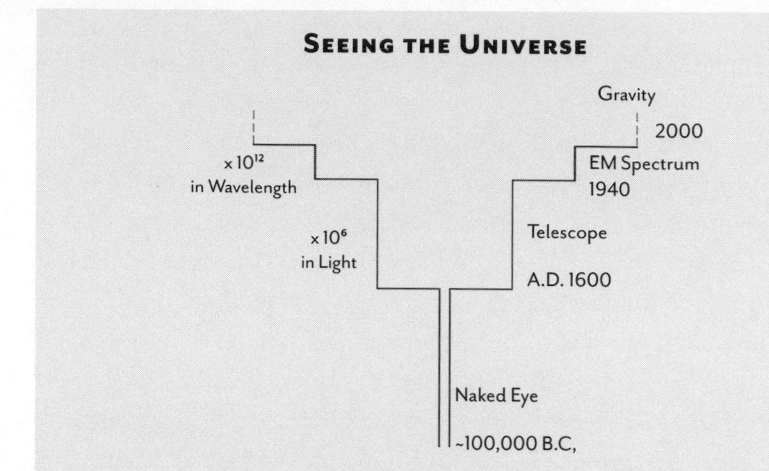

Figure 1.5. In the history of astronomy, huge gains in understanding have come with the invention of the telescope and the development of technologies to detect invisible waves. The detection of gravity waves lies ahead.

that describes how gravity curves space-time. He said that maybe ten thousand people have taken a graduate class in general relativity. But, he quickly added, that's not a good measure. "If you mean those who have fully absorbed its insights and elegance and can apply it in new situations, less than a hundred," he said. In other words, very few know relativity at gut level. The monks may never get into the Relativity Club, but in their intuitive grasp of large numbers associated with books, sand, and rice, they're experiencing the beating heart of science.

Bits of Information

In my university classes I use an audience response device to poll students on questions or their opinions. Each student holds a clicker, and in response to a prompt on a large screen at the front of the class they select from multiple-choice answers. We immediately see the class response in the form of a chart. Here in India, I've opted for the low-tech version of clickers.

The monks and I are about to explore the information revolution behind computer calculations, which is based on the bit, and the on/off switching system used to harness fast electronics into doing calculations and solving problems. For this lesson they'll respond to a question and possible answers I project onto a screen, and their "clickers" will be handheld sheets of paper. The monks have four answers to choose from, listed A to D. I give them each a piece of paper printed with A, B, C, and D in quadrants (and color coded as well), which they can hold up to signal their choices. With a couple of folds they can hold up just a quarter of the sheet with their preferred answer showing. We practice with a simple question, and once the monks have the knack of the system, we move to our central topic: how bits of information work.

I turn to a monk named Dawa Dorjee. "Dawa, think of a number from one to a thousand." Then I ask the card-wielding monks, "How many guesses would it take to get Dawa's number?" Before they reply, I display the four possible answers using our simple alphabetical code: A is 10, B is 100, C is 500, and D is 800. I give them a minute to think. Then I do a countdown with my fingers, ending with, "Now!" The monks hold up a sea of paper. The responses are equally divided between C and D, with a couple of monks choosing B. Nobody picks A. After all, who can imagine guessing Dawa's number in merely ten tries when there are a thousand possibilities? After a few minutes of discussion, there's general agreement that to guess the number in only a hundred tries would be exceedingly good luck.

So how many guesses does it take? We find out—the hard way. The patience of the monks makes this hands-on demonstration possible. A few of them go to the whiteboard and literally write out every number from one to one thousand. Once that's done, everyone sits. The monks begin to offer guesses. For ten minutes they call out numbers. Each time, Dawa shakes his head, and I cross the numbers off on the board. Finally, after 420 tries, someone guesses right: 178.

We try a more systematic approach for the next number that Dawa privately chooses: the monks call the numbers out in sequence and it quickly becomes a lilting chant. But this takes even more tries: 667 numbers are counted until we land on Dawa's pick. Discussing this tiresome exercise, the monks conclude that each of these guessing methods is equivalent. If the number is selected randomly, they think it will on average take 500 attempts to guess it. More than 500 is unlucky and less than 500 is lucky.

Among the "precept violations," or sins, that are forbidden in the Buddhist monastic tradition, gambling may not be the worst—but it's definitely on the list. I now delicately cross that line and tempt the monks to gamble. "I'm going to make you a bet," I say. When translated, this gets their attention. "What if I told you that I could always guess the number by asking ten questions?" They look dubious. Many shake their heads. I ask if they'd bet their burgundy robes against all the mutton momos they can eat on the wager that I won't succeed. Nobody speaks, but I can see I have their total attention. It seems that some of the monks, if they were betting men, would see it as a sucker bet in their favor.

With no takers I call upon a monk named Yeshi Choephel, a quiet older student who merits the title "geshe," and I ask him to think of a number. Now I begin my ten questions: Is it over 500? No. Is it over 250? Yes. Is it over 375? No. Is it over 312? Yes. I continue in this way four more times. My ninth question: Is it 343? He looks crestfallen. Yes.

My method is no secret, and it has nothing to do with luck, so I explain it to the monks and they try it out. The key is not to waste questions by guessing specific numbers. Imagine a number line from 1 to 1,000. The answer to the first question pins the number to either the low half or the high half of the range. Each succeeding question divides the remaining range in half. After ten questions, the mystery number is uniquely defined. This method is perfectly reliable and it works every time.

What the monks don't know up to this point is that we've been using the method of the "bit," which is a minimum unit of information. A bit is also known as a *binary digit*, because it can only be in two states: on/off, up/down, yes/no, loud/quiet, male/female, and so on. In a binary counting system, the only digits are 0 and 1. A single bit is the crudest kind of information. Adding more bits increases the detail and the descriptive power. For heat, one bit would be the distinction between hot and cold. Two bits of information about temperature would divide the scale into four parts: hot, warm, cool, and cold. For light intensity, one bit would be the distinction between black and white. Adding more bits of information would distinguish various shades of gray.

This reasoning applies to any quantity we can measure. For counting, one bit describes 2^1 or two numbers, two bits describes 2^2 or 4 numbers, three bits describes 2^3 or 8 numbers, and so on, up to ten bits which specifies 2^{10} or 1,024 numbers. That's how my "trick" works. Each question with a yes or no answer added one bit of information, and ten bits were enough to specify any number up to 1,000. The logic and the process can keep going. Twenty bits specifies any number up to 1 million, and thirty bits any number up to 1 billion. It occurs to me that the monks might have been willing to play the game up to a million the hard way. I smile to myself at the thought.

The bit is at the heart of information technology. Perhaps the monks who pursue this topic into higher studies will realize that the humble bit makes possible every call on a cell phone, every e-mail, every task run on a computer, and every word and image seen on TV or the Internet. They all result from a blazing-fast torrent of 1s and 0s. It's no shame to have overlooked the omnipresence of the bit, whether in Dharamsala or Arizona. American college students are digitally savvy as users but are on a par with the monks in being innocent of how their high-tech life works. Facebook might as well be magic.

I hold up a sheet of paper and make a dot on it with a pen. I ask half of the monks to split into their monastery groups to do the same thing. Their task is to see how many folds it takes to isolate the dot to within a square centimeter. I hand each group a pair of scissors. The other half of the monks also split into groups. I also ask them to calculate how many times they would have to subdivide the paper, but to use only mathematics to solve this problem.

Obviously, I've made it easier for one side of the room, the groups using paper and scissors. They make a dot on the sheet of paper, fold it in half, and then cut along the fold to discard the half without the dot. They fold the half-sheet of paper that remains, cut it again, and discard the half without the dot. They repeat this until they're folding and cutting bits of paper smaller than a stamp. Nine or ten sequential cuts are enough for them to get down to a piece of paper about a centimeter on a side. I peer over the shoulders of monks sitting cross-legged on pads on the floor, and several groups have gone beyond the call of the experiment. They hold up their index fingers with squares of paper adhering to them that are no bigger than the ink dots.

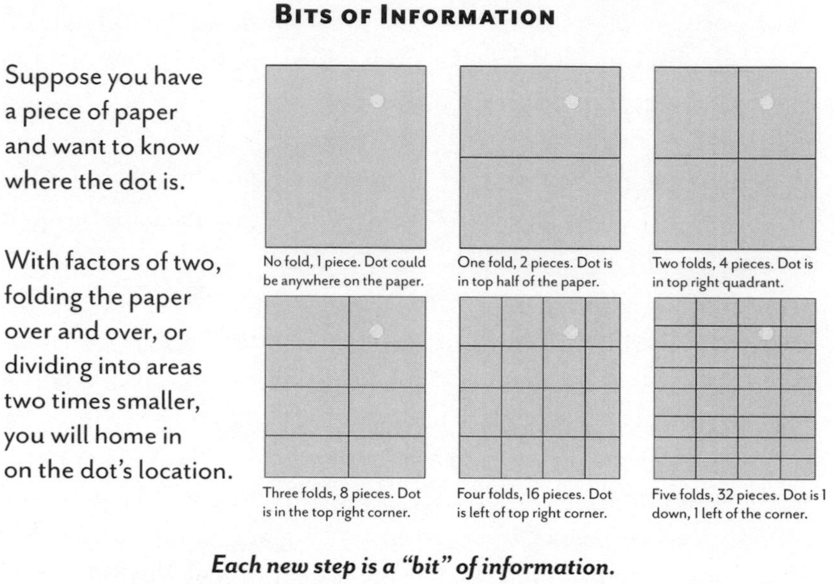

Bits of Information

Suppose you have a piece of paper and want to know where the dot is.

With factors of two, folding the paper over and over, or dividing into areas two times smaller, you will home in on the dot's location.

No fold, 1 piece. Dot could be anywhere on the paper.

One fold, 2 pieces. Dot is in top half of the paper.

Two folds, 4 pieces. Dot is in top right quadrant.

Three folds, 8 pieces. Dot is in the top right corner.

Four folds, 16 pieces. Dot is left of top right corner.

Five folds, 32 pieces. Dot is 1 down, 1 left of the corner.

Each new step is a "bit" of information.

Figure 1.6. Each question asked and answered about the natural world refines our knowledge. Any physical measurement can be boiled down to bits of information; more bits correspond to increased precision.

On the other side of the room, the monks are struggling to solve the problem with pure mathematics. Only one of the groups is close to the right answer. I call off the effort and instead begin a discussion of how it works.

The experiment is different from the number guessing, which would be analogous to defining positions in one dimension along a piece of string. The sheet of paper has two dimensions, so bits are needed to specify a postion in each dimension: five bits to define the width to a precision of about a centimeter and five bits to define the length to a precision of about a centimeter. This is because 2^5 is 32 so subdividing 30 centimeters five times reduces it to about one centimeter. Ten bits would define the position within a thousandth of the width of the sheet and ten bits for defining the position within a thousandth of the height of the sheet. So a process of twenty folds (or cuts) could narrow down the position of our dot down in each dimension by a factor of a thousand, which is less than a millimeter in each direction. But the monks quickly discover that it's physically impossible to fold a piece of paper in half and then in half again more than seven or eight times. Math is tough but brain beats brawn. Defining something with bits is very efficient.

Voila! It has taken only twenty bits of information to precisely locate the dot on a sheet of paper. Fifty bits would be enough to specify its position to the level of a molecule of cellulose.

For sheer fun—at least I'm trying for that—we extend the calculation to three dimensions. I get them to think about how many bits would specify the position of any molecule of air in the room. The mental exercise is similar to the paper folding; we imagine asking whether the molecule is in the back half of the room or the front half, the left side or the right side, or the top half or the bottom half. Each yes or no answer narrows down the volume by a factor of two.

This calculation is also a struggle for the monks. The numbers become very large very quickly. Still, they pull through. The room is about five meters on a side. A single nitrogen or oxygen molecule is about 5×10^{-11} meters on a side. That's a ratio of a trillion, which, in terms of divisions by a factor of two, would take 40 bits to specify. Each other dimension needs another 40 bits, so 120 bits are sufficient to specify the position of any molecule in the room. The monks have now gone beyond the call of duty by participating in these manual calculations. It has had unparalleled educational value. They deserve to see how bits work in an actual computer, which will provide a big dose of fun for them in the coming days.

I've brought to India a dome-shaped computer the size of half a honeydew melon. It's designed to simulate a classic guessing game. You may call it "twenty questions," but we'll just call this plastic gizmo the "mystery ball." With a flick on a switch at the base, it whirs into life. Red words project onto the inside of the dome and scroll from left to right. It says, "Do you want to play a game?" I hit the button labeled yes. "I can read your mind," it replies. The machine is very sure of itself. "Think of something, anything," it says. "I will guess it with no more than twenty questions."

The monks all crane forward to look. Who wants to go first, I ask? A monk named Jigme Gyatso shoots his hand into the air. He has thought of a cloud. As the machine asks questions by scrolling them across the dome, and as Jigme replies, I hit the appropriate button for his answer: yes, no, maybe, sometimes, or unknown. On the fifteenth guess, the machine stops beating around the bush: "Is it a cloud?" Jigme's eyebrows rise and his jaw drops. He's shocked.

Thupten goes next. The machine is even quicker. After eleven guesses it asks, "Is it a banana?" Three more monks fail to stump the machine. Yet this little computer isn't clairvoyant. Twenty yes-no questions yield twenty bits of information, enough to specify 2^{20}, or 1 million, items. Conceptually the guessing game is similar to our earlier exercises. Yet

This toy seems like it can read your mind. Think of anything you want, but try to avoid things that are very specific to your own culture or geography.

The game uses more than 20 bits because possible answers are:

YES
NO
MAYBE
SOMETIMES
UNKNOWN

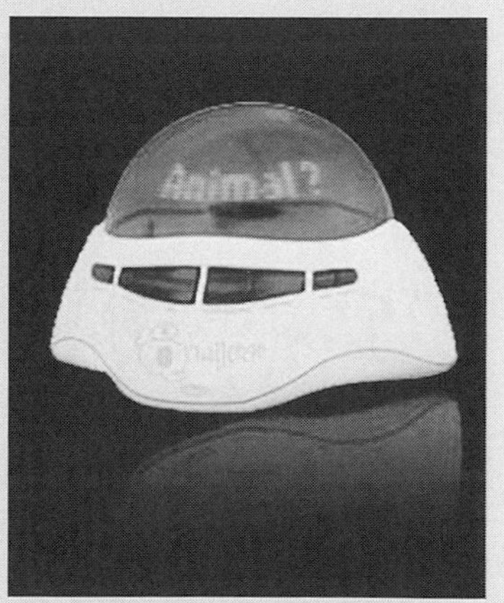

Figure 1.7. The 20Q toy is made by Radica, now owned by Mattel. It has a custom-designed chip to deliver and process answers to millions of questions. It narrows down and selects among the many options.

the computer can rarely be beat. Since most people can't think of a million different things and the programmers have anticipated and coded the most likely choices, the machine's odds of winning are excellent.

The Quantum World

The next morning in class we discuss one of the milestones in modern physics. It's the revolution from classical physics, which describes one kind of reality, to the realm of quantum mechanics, another reality entirely. Classical physics began with Isaac Newton, who formulated one law that explained gravity, and thus the motion of objects on Earth and in space. Science is rooted in the search for cause and effect, and so Newton's theory seemed to offer a very deterministic world, which some of his followers would call a "clockwork universe." This concept worried a lot of people because it seemed to threaten the notion of free will.

That worry was misplaced, however, because the real world is more complicated than the simple physical description that Newton's law of gravity offered. Newton's law refers only to the force between two objects. In any realistic astronomical situation, such as the solar system or a star cluster or a galaxy, many bodies are moving and attracting each other. Even a powerful computer can't calculate with absolute accuracy how all the bodies determine each other's movements and fates.

Nevertheless, Newton's simple, deterministic law was so successful that physicists were unprepared for its partial overthrow in the 1920s. At that time physicists working on a microscopic theory of nature found that atoms and particles weren't finite entities. For example, particles didn't have sharp edges but could behave like waves. Also, they didn't operate like clockwork but could behave unpredictably. Newton's determinism gave way to "uncertainty" in physical laws. The result: science faces a fundamental limit to the precision of any measurement. Nature is fuzzy, not sharp, and it's probabilistic, not deterministic.

This approach is called "quantum" theory for the way particles now seem to move between different states by a discrete leap called a quantum. Quantum theory was, and still is, shocking and counterintuitive. Niels Bohr, the architect of the modern theory of the atom, said, "Everything we call real is made of things that cannot be regarded as real." John Wheeler, a leading American physicist, stated, "If you are not completely confused by quantum mechanics, you do not understand it." So disruptive was quantum theory that Albert Einstein fervently hoped that it was wrong. He wrote gloomily, "If it's correct, it signifies the end of physics as a science."

Even mortals can play around with quantum theory, and I give the assignment to the monk Jigme Gyatso, who had cleverly thought of a cloud for the twenty-questions game we played earlier. I position him in the corridor outside the classroom, clutching a rupee coin. These coins can be hard to find in northern India, since the steel it's made of is worth more than one rupee. Some people even melt them down to make razor blades of higher value. I've asked Jigme to take his coin and spin it on the tile floor in the hallway so that it finally lands on heads or tails. He is to do this a minute after I leave him, and then he'll wait five minutes before bringing us the results.

Back in the classroom, all the monks agree that the coin has equal odds of ending up as heads or tails. Then we hear the sound of the coin spinning. I ask the monks, what's the state of the coin? It has no state, many of them say, it's still spinning. Then we hear the coin clatter to a stop. Now what's the state of the coin? Heads, some shout out. Others say tails. I tell the monks they're just guessing. What's the most rational way to describe the coin? They discuss this in their monastery groups, and I walk among them, sampling discussions that are punctuated by hand claps and gusts of laughter.

Most groups decide the question can't be answered, even after the coin stops spinning. But, as I point out, the question *does* have an answer. Jigme knows whether the coin is heads or tails, and so does anyone who happens to walk past. It's just *we* who are in a state of ignorance. With that assertion, a few groups of monks say that such ignorance is fine; they can accept that the coin is in an indeterminate state. Then Jigme comes back in. "It landed heads," he says. Now that we all know, the uncertainty has evaporated.

This, in a nutshell, is the weirdness of quantum theory. The world of subatomic particles is one where behavior is governed by probability. The spinning coin shares two equally probable states. It's not heads and it's not tails—it's both. Even after it stops spinning, we have no better description of its state. Only when we make the observation, through Jigme or by going out into the corridor, does the uncertainty disappear.

The key point in quantum theory is that the moment of observation changes the quantum reality, and you get one state only. Technically speaking, the observation causes the wave function of the particle to "collapse," becoming, in effect, heads or tails. Put another way, whatever nature truly is, we make it a reality by our observation only. Einstein found this situation abhorrent. He imagined a deeper description of reality that wasn't affected by our observation and that we could "know" that description, like Jigme "knowing" the state of the coin when it stopped spinning (even though those of us in the room knew only probabilities). After nearly a century of testing, quantum theory has proven to be a robust way to describe the behavior of subatomic particles. It's the only way we have to describe microscopic reality. We're stuck with its weirdness.

Another foundation stone of quantum theory is Heisenberg's uncertainty principle. I caution the monks that the spinning coin is an analogy, and all human-scale analogies fail to capture the subtlety of quantum theory. It turns out we can't "know" the state of any particle with absolute precision. Heisenberg's principle says that when interrogating the state of a particle—which has velocity and position—we can't measure both at the same time with absolute precision. Knowing the velocity well means we know the position poorly, and vice versa. The accuracy of our measurements has a limit. We're not uncertain because of bad equipment or incompetence of the experimenter, but because nature presents us with a fundamental barrier.

To make this quantum world real, we clear the floor. Jigme stands in the middle of it. I switch off the lights and shine a flashlight around

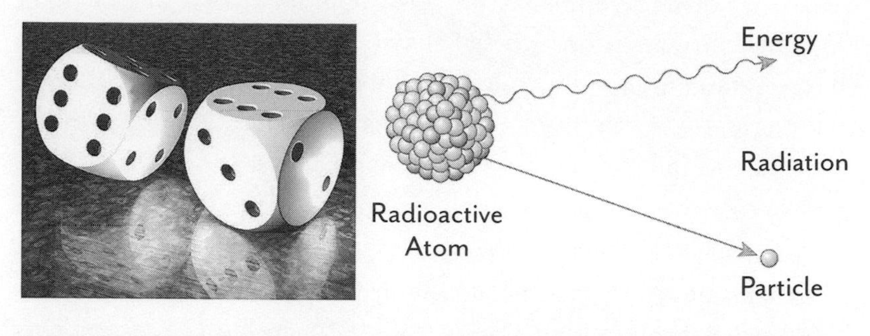

Figure 1.8. The microscopic world is governed by indeterminacy, and Heisenberg's uncertainty principle sets a limit to knowledge. But the behavior of a large ensemble of particles or objects is repeatable and predictable.

the room. It catches him; he flinches and blinks in the beam. "This is how we locate 'Jigme the monk,'" I say. "Light bounces off him, and we see where he is." In reflecting off him, the light does actually transfer momentum to him and give him a "push" but it's a tiny effect, too small to measure.

Now we shrink to quantum scale. Jigme is still the object we want to locate but now he's "Jigme the particle." Three other monks stand on the side of the room blindfolded, holding foam balls. The foam balls are photons—particles of light. If they want to know where Jigme is, they have to shine a light on him, which means throwing foam balls at him. I tell him to shout if a foam ball hits him. The ball will give him energy, and he'll move in an appropriate direction. If the foam ball hits him head-on, he should move backward, and if it glances off him on either side, he should move backward and sideways, like a pool ball. With the rules established, the foam balls start flying.

The first dozen miss and the rest of the monks are greatly amused. Then a ball hits him on the shoulder and he shouts, "I'm hit!" So where is Jigme? I ask the monk who threw the ball. He's blindfolded, but he points in the direction of the shout. The other monks laugh, because Jigme has in fact ricocheted to the other side of the room. The act of observing Jigme's position has given him motion, which leaves him in a different place. We can't know about something without interacting with it. Quantum indeterminacy casts a veil over our knowledge of the world.

We've encountered one of the dualities that will arise often in our journey through science and Buddhist philosophy, I say to the monks. We'll explore the large and the small and the duality of nothing and something. In this case, we've pitted everyday reality against quantum reality. Our normal world of large objects is smooth and measurable. At the subatomic level, quantum properties are discrete and seemingly unknowable. We move through our lives oblivious to the evanescent and shifting landscape of subatomic particles. I hear murmurs of approval. The idea of impermanence sits well with Buddhist monks.

That night in our little enclave down the slopes from Dharamsala, a crescent moon casts pale yellow light on the fields and rice paddies to the south. It also hits the front edge of the Himalayan range to the north. The stars twinkle through the humid air. Paul and I are relishing the sensory experience; the air here is fragrant with mimosa, sandalwood, and cumin. There's a rustle of insects and the call of night birds in every direction. Neither of us is sleepy, so we go to the rooftop of the dormitory building to survey this sumptuous tableau. We keep away from the edge since there's no railing and it's a sixty-foot drop.

Above us is a familiar sky. To see it back in Arizona I can drive from the desert floor to a silver-domed observatory on a mountaintop, a complex with millions of dollars' worth of high-tech equipment. Here on the rooftop, I'm reduced to my essence as a naked-eye observer. Yet it's the same sky everywhere. The stars seem to move, but really they just pass from one country to another and from one tribe to another as the planet turns. Paul and I linger in silence and reflection.

Like stars, people can change locations and yet stubbornly stay the same. Paul and I have traveled to a new place believing, perhaps, that we will experience a magical change, but in reality, we take who we are with us—often the revelation of travelers and wanderers. We're hoping to change our state of mind after my recent divorce. Our family dissolved after twenty years of marriage, and I know it has taken a toll on my son. Now we've come to an unfamiliar world and we're waiting to see what happens to our inner lives. Paul hasn't been anywhere more exotic than Italy. I'm still worrying about how he'll fare and what he'll do while I'm teaching. But we've decided. We've embarked on an adventure that lets us escape the routines of our daily lives and regroup.

There's more. Being in a strange place is loosening chunks of introspection, like an ice floe breaking up. I'm here to teach science, and I'm convinced of its essential importance and validity. I'm personally invested too—my career and status are tightly bound to the pursuit of knowledge. But I've just been talking about the primacy of doubt. How do I square this with my expertise as a scientist and my authority as a teacher? If I'm being totally honest, it also relates to the high value I place in knowing something, and in not being wrong, which is a personal failing. Can I really be humble before the void?

These thoughts are personal and existential. But among the monks they can become cosmological as well. We can compare our individual fates to the much larger processes of the universe, and even to "reality" itself. A few hours earlier the monks and I took our science discussion into this realm of philosophical reflection. We talked about uncertainty as a core truth of reality. Acceptance of uncertainty is part of the Buddhist outlook, but it has its parallel in physics. Some of the best scientists have embraced the tension between knowledge and uncertainty. As James Glieck says in his biography of Richard Feynmann, "He believed in the primacy of doubt, not as a blemish on our ability to know, but as the essence of knowing." Buddhist tradition accords with this perspective. The humility and playfulness of the monks suggest they don't need final conclusions. But they still like to talk about "what's real."

Soon after I arrived, Geshe Lhakdor, head of the Tibetan Library, offered me this pithy definition of reality: "The core of Buddhist thought is that nothing has independent, permanent, or absolute existence." The universe is a limitless web of interconnections and undergoes a continual process of transformation. Everything affects everything else. This law of mutual causation, so complex that it remains elusive to our human apprehension, is what Buddhism calls karma. And it sounds a lot like the quantum mechanics we discussed earlier in class.

Today, quantum mechanics has forced many scientists to give up the old view of reality as material and recognize that what seems concrete to us—matter and consciousness—results from interactions of matter and energy. The idea began in ancient Greece with Thales and Anaxagoras, and it entered Buddhist tradition soon afterward. The cofounder of quantum theory, Werner Heisenberg, offered this summation of the elusive subatomic world: "The ontology of materialism rested upon the illusion that the kind of existence, the direct 'actuality' of the world around us, can be extrapolated into the atomic range. This extrapolation, however, is impossible. . . . Atoms are not things."

Echoing Heisenberg's outlook, many scientists now prefer to speak of "naturalism," which is a broader philosophical position than old-fashioned materialism. Naturalism holds that everything in the universe can be explained by natural causes. It has no place for supernatural phenomena such as miracles. Going further, naturalism also supposes that the universe is devoid of purpose and indifferent to human desires and needs. Naturalism squeezes the magic and mystery out of life. And in a sense, that's the whole point. There must be a natural cause, science declares, even if it isn't discernible. This assumption has been borne out by experience, but it must be constantly tested. I find naturalism an optimistic worldview because it says we're not subject to mysterious forces, and it assigns us the power of understanding and then controlling our own destiny.

Similar to naturalism is the Buddhist concept of "dependent origination." According to the Dalai Lama, two kinds of viewpoints must be rejected: "[First] that things can arise from nowhere, with no causes and conditions, and the second is that things can arise on account of a transcendent designer or creator." While very few people go around saying that nothing causes everything, many people seem to favor a Creator. On this, I agree with the Dalai Lama's view, though I call myself an agnostic on such metaphysical questions. But like most human beings

I can tremble in the face of a universe that's ultimately transient and meaningless. Buddhism goes even further than quantum mechanics, stating that reality itself is an illusion. "So, reality is an illusion," I say to myself. I try to find the benefits in such a view. What solace does it give me at a time of difficult personal relations? The bruised feelings of my heart and my concern for my son don't seem illusory.

Short of an answer, we all—myself, the monks, and even Paul—throw ourselves into activity in the hope that learning will be a balm. We did this in our last class by putting aside the "mystery ball" and creating a human guessing game in the spirit of the American physicist John Wheeler. He spent a lifetime trying to understand—or explain in simple stories—how quantum reality works. We played Wheeler's game like this: We have chosen Nyima Gyaltsen to be our twenty-questions computer, so to speak. He leaves the room believing we've thought of a "thing" that he has twenty questions to guess. However, when Nyima returns we really have *not* thought of a thing. Each monk is to instead think of any "thing" he wants as Nyima asks his questions. The game is challenging because each successive monk has to think of a thing consistent with all the questions and answers that have come before. In effect, everything is contingent.

After eleven questions, receiving yes and no replies, Nyima makes his guess. The room explodes in laughter. The item Nyima declares into existence had no prior existence. Instead it emerged in response to the questions asked. Its nature depended not only on the questioner but on the responses. As a quantum physicist, Wheeler saw this version of the game as symbolic of the strangeness of quantum theory. We live in a world where no phenomenon exists independent of the act of observation. With our game, the monks and I have felt the thrill of cocreating reality.

CHAPTER 2

Scales of Space

ANYTHING CAN HAPPEN in a roomful of Tibetan monks. The same is true at a breakfast table with just one senior monk. At our morning meal the program staff, including the Tibetan translators, is discussing the best topping for the daily staple, a doughy substance halfway between pancake and naan. Some favor mango jam. Others like peanut butter, and a few argue for soft cheese. After selecting his taste preference, Geshe Lhakdor, head of the Tibetan Library, waxes philosophical. He muses on the difference between sensory reality and ultimate reality.

"There's the table and there are the chairs," he says. "But we each have our own experience of them, an experience not shared. Beyond all of that, however, is a reality that Buddhism calls 'not table' and 'not chairs.' The Buddha said no seeing is seeing." At this, Geshe Lhakdor offers a broad grin, a gesture telling us that Buddhists confront such ineffable matters on a daily basis—but never take them too seriously.

On another occasion, a monk asks me with great earnestness, "What is the nature of space?" I've assured the monks that there's no such thing as a stupid question. Today we'll discuss the topic of space, within astronomical limits. But these philosopher-monks want to know more. If invisible space didn't have stars and galaxies in it to mark it, could we measure it at all? What is space if it's really nothing, and how can nothing expand? Is space quantized like matter, or can it be infinitely subdivided? Is it a phenomenon or just an abstraction?

These are among the great metaphysical questions of the cosmos, and I will have no answers for my students. As they ask, I feel like a child in the garden of the universe. We proceed with what we know: the size and volume of the space that astronomers can detect.

The Size of Space

Space is absolutely huge and that gives cosmology its unique character: It's always about huge numbers. To come to terms with hugeness, the monks need a basic skill of estimation, or how to come up with a ballpark number. Very few astronomical numbers are known with a precision better than a factor of two. Therefore, in the first exploration of a topic, it's useful to come up with an order-of-magnitude estimate, or a number within a factor of ten of what you'd get from a detailed calculation. This seems hopelessly crude. Imagine if you asked a real estate agent to estimate a sale price for your house, and he says, "I think it's worth $400,000, but it could be $4 million and it might be as low as $40,000." You'd probably find a better agent.

But an estimate isn't the same as a guess. If the starting point is an unfamiliar situation, any handle on a number is useful. Estimation is done with sufficient care that it can't be totally off. In a previous class, the monks said the largest numbers that are part of their experience are about a million. Then we reached a billion by estimating the number of grains of sand in a bucket. It turns out that the monks' tradition contains a truly prodigious number. In a text attributed to the Buddha from the first century BCE, he competes with five other suitors for the hand of a princess. After defeating them all in writing, wrestling, archery, running, and swimming, he's given a final test by the eminent mathematician Arjuna: "Young man, do you know the counting that goes beyond the koti [10 million]?" Buddha ascends the ladder of numbers and doesn't stop until he reaches the gleaming pinnacle of 10^{421}, one followed by 421 zeros.

There's no number in science as large as 10^{421}, but thanks to this Buddhist tale the monks have had exposure to ridiculously large numbers. Today, this will allow us to construct the universe in stages.

As always, Thupten B is willing to assist. I ask him to be the main

A Scale Model

=

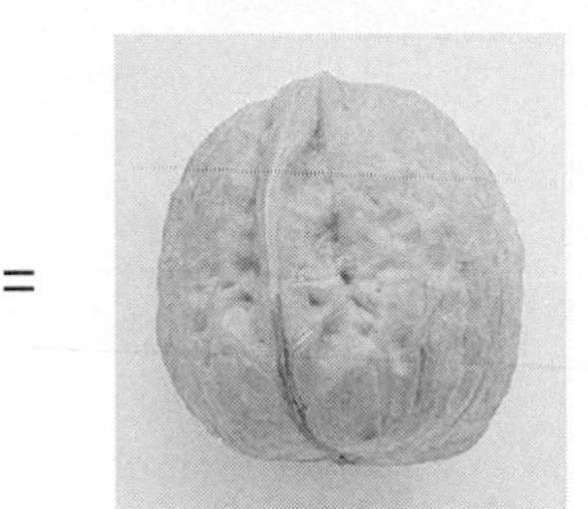

Set the Earth to the size of a walnut, or a 1:300,000,000 scale model.

- The Moon is a pea at arm's length.
- The Sun is a 3 m ball 100 m away.
- Neptune is another pea 2 km away.
- The nearest star is 50,000 km away.

Figure 2.1. In this scale model of the solar system, the planets fit within a region the size of a small town, but the nearest star doesn't even fit on the Earth. At this scale, light would travel at walking speed.

prop for our scale model of the universe. We begin with a miniature Earth I have in my teaching bag of tricks; it's made of metal and is the size of a small lime. Thupten holds it between his thumb and forefinger. Then I hand him a small, milky-white marble that stands for the Moon and challenge everyone to decide its proper distance from the model Earth. They mostly think the Moon snuggles close to the Earth, and in Thupten's first estimation, he has the Earth and marble resting on either cheek, with a grin between them. I take his hand that holds the marble and pull his arm out full length. That's the distance to the Moon and it's the farthest humans have ventured. When I explain that the Apollo moon shots cost $50 billion in inflation-corrected dollars, there's a bit of shock. Later at lunch, Gelek Gyaltsen and Thupten talk about being puzzled at spending such a vast sum of money out in space in a world with so many human problems.

The building of our model continues. A tall, thin monk named Lobsang Choephel comes forward to be Mars, and I hand him a larger marble, a red cat's eye. The class thinks Lobsang should stand on the far side of the classroom. They laugh when I direct him out the door to the far side of the building. We see him through the window, standing across the courtyard holding his treasure, the miniature god of war. On this scale, I explain, the Sun is a three-meter globe five hundred meters away. I need another volunteer and Geshe Nyima obliges, nimbly rising from a cross-legged position. I hand him a pale blue plastic ball the size of a grapefruit. It represents the outermost planet, Neptune. With a straight face I ask him to walk up to McLeod Ganj, the upper town of Dharamsala. It's twenty kilometers away. He plays along, sighing and marching out the door—until I call him back.

What we've created, I explain, is an orrery: a small mechanical device that displays the motions of the planets in the solar system. In our case, it is a monk orrery. But with sizes and distances on the correct scale, our model is too dispersed, so we shrink it into the classroom. As everyone clears space in the room, we look at how orbits work.

The hub is the Sun, a role I give to Geshe Nyima, whose girth and warm smile, and the owlish way he wears his glasses, seem just right for our Sun god. Like all the monks, Nyima has a story. He arrived in India as a teenager in exile. The school he attended in Delhi wasn't rigorous enough, so he traveled to Madras to attend a small Catholic school. I try to imagine a young Tibetan kid living with Catholic nuns while surrounded by Hindus. He's not to be underestimated: Nyima speaks Tibetan, Chinese, Tamil, Hindi, and English, and this facility has made him one of our translators. He eats with the teachers and stays in our lodging, increasing the opportunities for us to hear monk biographies.

Now in the Sun's position, Nyima is our epicenter for the planets. As the Earth, Thupten stands three meters away, and by shrinking one year to a minute, he walks around Nyima once in sixty seconds. Monk Mercury joins in, pacing one meter from the Sun and taking only fifteen seconds to circle around. Monk Venus is two meters away, orbiting in twenty-five seconds. Monk Mars is four and a half meters out; his orbit finishes in under two minutes. The rest of the monks stand at the perimeter of the room to form the constellations of the zodiac, a circle of animals.

Go! The monks are in motion and adjustments are quickly needed. Mercury must face the Sun as he orbits due to tidal locking, as must the

Moon. Tidal locking occurs when a small body orbits close to a larger one; gravity grips the small body so that it keeps the same face pointing at the larger body. The choreography is challenging. I've picked Dawa Dorjee to be the Moon since he's barely five feet tall and nimble, but he struggles to pivot around the moving Earth every five seconds. There's loud whooping and chortling from the zodiac as he scrambles. Mercifully, I've relieved Thupten of the Earth's onus of spinning six times a second to show day and night.

Then we stop. It's time to contemplate the empty space that fills the solar system. The Sun comprises almost all of the mass and gravitational force, and the eight planets are negligible in comparison. The four rocky planets are the size of cherries and plums, and they orbit within a half mile of the ten-foot-diameter Sun. Farther out are four planets that are actually large gassy spheres, and in this model the size of baseballs and beach balls. Apart from a swathe of orbiting sand particles to represent asteroids, and occasional dust-mote interlopers to represent meteors and comets, this is all the material in the solar system. Several monks are curious about Pluto's demotion from the pantheon of planets. But they show none of the dismay I encounter with public audiences back home. They're just not that attached.

If the solar system is the size of a large city, the nearest stars would be nowhere on Earth. Now we're getting a perspective on the challenge of space travel. The cost of going to the Moon was so great that we've not been back for over forty years. The nearest star, Proxima Centauri, is 100 million times farther away than the Moon, beyond our wildest dreams.

Our large-city scale for the solar system, which shrinks the universe 300 million times, also allows us to visualize the speed of light. Its speed shrinks by the same factor, from a blistering three hundred thousand kilometers per second to a sedate one meter per second. Radiation travels at a brisk walking pace across that large city. I ask the monks to imagine themselves as information carriers, or photons, in this scale model. They can see it would take eight minutes to saunter here from the Sun. They could cross the solar system in five hours. But it would take several years to walk here from even the nearest stars.

This scale model barely captures the distances between stars, and it reveals nothing about the Milky Way. So we shrink the scale by another factor of 100,000,000. I reach into the bucket of sand that's been sitting in the corner of the classroom. Relieving Thupten of his Earth and Moon, I place a tiny grain on the tip of his forefinger. Be careful, I say, that's the whole solar system. Six billion souls depend on you to not drop it. He accepts the responsibility with equanimity. I take the Mars marble from Lobsang and put a grain on his finger as well, then position him halfway across the room. You're the nearest star system, I tell him: Proxima Centauri, which has two stars orbiting each other. In fact, I tell the monks, each of its Sun-like stars may harbor Earthlike planets. Lobsang has been daydreaming and gazing around the room, but when he hears this he perks up and stares thoughtfully at the speck on the tip of his finger. Who knows? It might harbors billions of sentient creatures of unknown function and form.

Reduce the scale by a factor of 100,000,000.

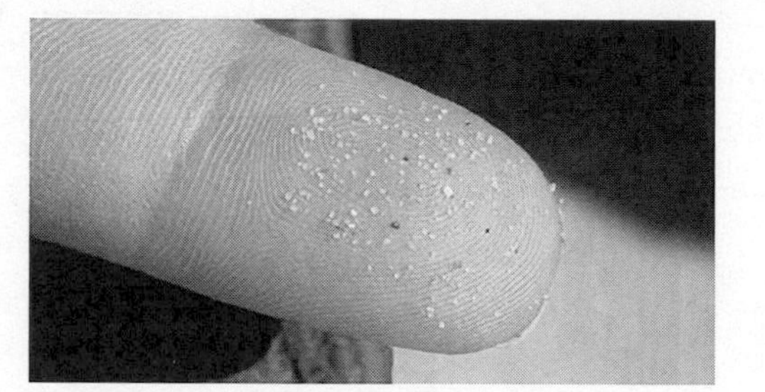

- The solar system is a grain of sand.
- The distance between stars is 10 m.
- The Milky Way is the size of India.
- The Milky Way has 400 billion stars.

Figure 2.2. After a second stage of scale reduction, the full extent of the solar system fits within a grain of sand. Stars are widely dispersed in the Milky Way and so almost never come into contact or collide.

Even by reducing stars and planets into the size of sand grains, we've still not been able to fit a single galaxy into the classroom. If we were to put sand grains (which stand for solar systems) a few meters apart, then the Milky Way would be an agglomeration of 400 billion grains covering the area of India. We'll have to scale things down ever further to even approach the entire size of the known universe. So we shrink a spiral galaxy like the Milky Way to the size of a dinner plate. I happen to have a stack of dinner plates in my bag since I borrowed them from the kitchen last night. We push aside the mats to create a three-dimensional model of the nearby universe.

Everyone gets to be a galaxy. With some careful planning I can use the tiles on the floor as an X-Y coordinate grid and the height above the floor as the third Z coordinate. I've figured out the positions ahead of time with a calculation done on my laptop, so the monks can count out distances to any galaxy in what astronomers call our Local Group of galaxies. A lucky few monks get to be the substantial galaxies: the Milky Way, M33, and M31 (also known as Andromeda). After some discussion the three dinner plates take their places. Now I produce a box of cotton balls, and I ask a monk to hold two wads of cotton a foot or so from the Milky Way. These represent the Magellanic Clouds. They can only be seen from the southern hemisphere, appearing as fuzzy aggregations of stars, gas, and dust. Two dozen more dwarf galaxies take their places around the room, held by monks near the floor or above their heads or, in a few convenient cases, resting on their heads. Our classroom is now a million light-years side to side.

So far the numbers sound pretty big, the monks agree. However, we are still talking small potatoes. A million light-years is a tiny fraction of the universe. If we could extend our scale model, we would place dinner plates across the campus grounds and out into the countryside and also high in the air since they occupy three-dimensional space. We'd also need to include small galaxies, suspending cotton wool balls beside and between all the dinner plates, like tiny clouds. Indeed, we'd need an armada of trucks filled with plates because we'd be doing this until we'd placed 100 billion galaxies across the entire Indian subcontinent. Only then would we have represented the visible universe. I imagine that Indian families might do this by flinging their crockery about, perhaps in a fit of pique at the enormity of the universe.

The monks enjoy our mind-expanding exercise. But I wonder how effective it really is. This self-doubt of the modern educator persists: How do you know that true learning, let alone life-changing insight, is taking place? From one point of view, the models and analogies we used are merely crutches, substitutes for mathematical descriptions of physical reality. Nevertheless, they offer a way to grasp the unthinkable size of the universe. It turns out that Buddhists have their own analogy-as-a-crutch, but it's more poetic and metaphysical than our dinner plates. Traditional Buddhist cosmology posits Mount Meru as the center of the physical and spiritual universe. Home to Lord Brahma and the demigods, Mount Meru is a million kilometers high, or a hundred times the size of the Earth. Besides this example, ancient texts don't assign sizes to larger realms. There's nothing in Buddhist training to prepare for the scales we're dealing with now. I wonder how many of the monks retain a belief in the Mount Mehru model.

Now reduce by another factor of 100,000,000.

- The Milky Way is the size of a plate.
- The nearest galaxy is 10 meters away.
- The universe is the size of India.
- Billions of galaxies fit within this space.

Figure 2.3. After twenty-four orders of magnitude reduction from their real size, spiral galaxies are about the size of dinner plates and the observable universe spans a distance of thousands of miles.

Teaching by analogy is only one dilemma for the modern educator. We also face a glaring fact of modern culture: short attention spans.

Research shows that the attention of any audience drops by 50 percent after fifteen minutes of lecturing, no matter what the topic, no matter how dynamic the lecturer.

Defying this research, the Tibetan monks are a delight. Their stamina is phenomenal. Their attention always seems fresh, perhaps the result of the years of mental exercises in concentration, patience, and meditation. The downside is their tradition of often taking in information from a master teacher without analysis or asking critical questions. The word of masters and gurus is unquestioned. The price of the monks' respect for me as a cosmology "geshe" might be uncritical acceptance of everything I say. If they're going to master cosmology and teach and debate the ideas with other monks, they'll have to poke and prod the information, testing its limits.

In our classroom, at least they feel free to ask questions. After three hours of teaching on the size of the universe, I'm drained, but the monks are coming to life. They pummel me with questions. How do we know how many galaxies are out there? What about the edge? What lies beyond the last galaxy? How do we measure sizes this large?

Every day so far, I've been conducting an internal monologue. What works best in the classroom? Anything that lets them be active learners. Is it question-and-answer? Yes, but they still need to get the information. Do I lecture too much? Lecturing works, but only in small doses. Am I mistaking attention for comprehension? I'll only know that if I test their knowledge. My goal is to see the monks interacting as much as possible, but meanwhile, given what I know of Buddhist cosmology, I need to shift their entire worldview. This can't be accomplished passively.

Still unresolved on the perfect learning model, I know that one feature of our program is turning out to be very beneficial. We've structured the classes to incorporate more time for reflection and digestion. Even peer learning—novices in science teaching other novices, occasionally incorrectly—has some benefit for the monks. So do the "Burning Question" sheets I make and circulate: monks write down questions as they arise, and I answer them after they get translated during a break.

Our educational approach is often a topic among the Western staff at breakfast, with Bryce acting as the genial emcee. The program designers and analysts, Richard and Mark, have been working with the monks on their writing and journal-keeping. To do this the monks gather at evening sessions. As requested by the Dalai Lama, the monks should emerge as able teachers to instruct others in the sciences. Richard and Mark are playing a big role to make sure this kind of progress takes place. Teaching science in a monastery setting is a new horizon for all of us. Western educators are figuring out how best to structure learning activities as the many workshops, mine included, unfold each season and year by year. The Tibetan staff is invaluable in facilitating the learning, from the total engagement of someone such as Thupten B—who I once overhear telling his group it's okay to question my assertions—to the skill of our four translators, who rotate in that capacity to avoid burnout.

Cosmic Perspective

Buddhist monks are not very worldly. That's part of what makes them special. Most have cell phones and e-mail addresses with Yahoo or Google, but they don't watch TV and they don't read newspapers. Some are on Facebook but the frenetic demands of social media have passed them by. Some have traveled within India, but few have been to any other country. If they've been abroad, it's to Tibet. Steeped in traditional texts, they live plain lives dedicated to prayer, study, meditation, and serving others.

All of this leads me to call them "otherworldly," but I shouldn't be too casual in my presumptions. Their grasp on the physical world is evident, or else we wouldn't have a means in class to imagine the cosmos. Like all of us, that grasp begins with childhood and memories of physical places. When any of us were four years old, for instance, how did we know our world? It probably began with intimate knowledge of our house, and then the street we lived on. Later it expanded to landmarks in the hometown. Beyond that, locations were just abstractions, names on a map. Flying over cites, provinces, or nations in an airplane didn't register them as real physical spaces.

So it is with the universe in the childhood of our understanding. We've gotten to know much about the surface of the Earth, but after that, we've only stepped into a neighbor's backyard (the Moon) and taken pictures of the other side of the city or mapped other towns (the planets). We may be familiar with the nearest thousand stars or so, but the number of galaxies is so stupendous that our neighborhood and town analogy breaks down. There are 400 billion stars in the Milky Way galaxy and 100 billion galaxies in the universe. The 7 billion people we share the planet with are abstract to us, so we can't truly imagine 100 billion objects—whether we're worldly Americans or otherworldly monks.

But today we try, returning to our ABCD card system to solicit responses from the monks as they try to take in the sheer scale of the cosmic perspective. In this round the responses are tentative. The monks look sheepish as they hold up their answers, not wanting to contradict their peers. Clearly, this is their first class in ordering a gigantic physical universe. Many still think we live on a moon orbiting a planet. Less than a third give the correct order of astronomical terms in decreasing order of size: universe, galaxy, nebula, star, and planet. We've got some work to do. Science shouldn't be obscured by jargon, but terminology is needed for the monks to engage in any meaningful dialogue with me and each other about the universe.

On top of the technical terms, the Earth has a philosophical address: "Nothing Special." This realization began when Nicholas Copernicus used logic and mathematics to make the case that the Earth isn't the centerpiece of the heavenly bodies (but rather the Sun, at least for the time being). The Copernican Revolution has expanded over a half-millennium, showing that our Sun is not the center of the universe, nor our galaxy, and the very presumption of a center is flawed. After reviewing the story of Copernicus, I tell the story of a successor in the Earth-location debate. We meet the wild-eyed mystic Giordano Bruno, who went far beyond Copernicus by speculating that the stars were remote suns with planets around them, and perhaps with familiar forms of life inhabiting those planets. The monks smile and nudge each other when I show a slide of a statue of Bruno, a robed and cowled

The Copernican Revolution

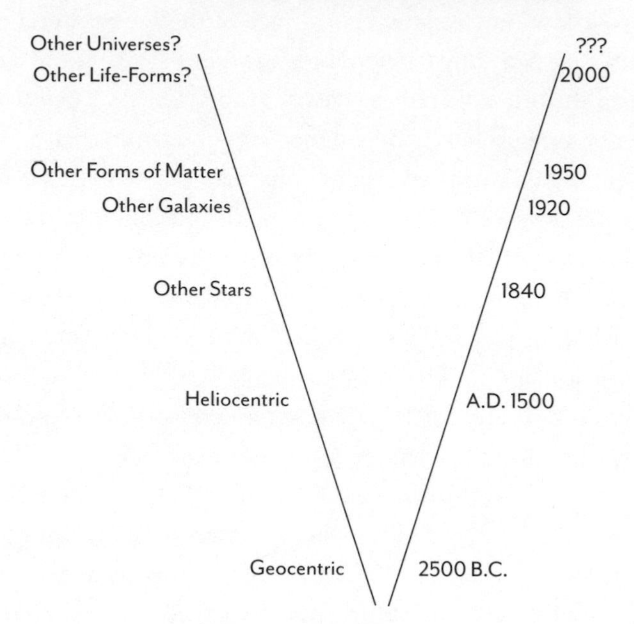

The history of astronomy displaces us from cosmic importance.

Figure 2.4. The Earth isn't special in terms of the star it orbits or the galaxy it inhabits and perhaps also in terms of the life it harbors. We're not even made of the type of "stuff" that constitutes most of the universe.

cleric, standing in the Campo de' Fiori in Rome. They become somber when I tell them that Bruno was burned there at the stake for his heretical views.

Gelek's hand goes up. "What did Bruno do wrong?"

At the Inquisition, I explain, Bruno was charged as an "impenitent and pertinacious heretic." It wasn't his astronomy that got him into trouble, but the fact that he questioned core premises of the Catholic faith. I'm sure the monks are puzzled.

In Buddhism, there's no heresy or blasphemy, and there are no heretics or infidels. Of course there has to be a boundary between Buddhism and non-Buddhism or sorta Buddhism. That usually comes from belief in four premises called the "dharma seals." These premises are that all compounded things are impermanent, all stained emotions are painful, all phenomena are empty, and nirvana is peace. But when a self-proclaimed Buddhist chooses to believe in an essential soul or that some things are permanent, nobody will stretch that person on a rack. Rather, the historical Buddha urged people not to form their beliefs through speculation and not to grasp at explanations that come from their imaginations. He said we should use observation and insight instead of forming a belief system.

That's what Copernicus did with his insights, and the Copernican Revolution continued to advance in the eighteenth and early nineteenth centuries. That was when William Herschel mapped out the disk of the Milky Way and other astronomers succeeded in measuring the distances to the nearest stars. Our solar system was getting smaller in importance as we realized it was adrift in a void, swirling along a lazy quarter-billion-year orbit of a vast system of stars. A little less than a century ago, Edwin Hubble demonstrated that many faint smudges of light in the sky were remote systems of stars, or "island universes." Since then, larger and larger telescopes have gathered ever more distant light to take the census of galaxies to a level 10 billion times fainter than the eye can see.

Yet most people still cling to Mother Earth as special, even though it's only a piece of rocky debris left over from the birth of a middleweight star. Our solar system is in a fairly quiet suburb, far from the bustle of the galactic center, between the thoroughfares of the Orion and Perseus spiral arms. If we could rise above the Milky Way, the Earth wouldn't catch our attention at all. And the Milky Way is not the fairest galaxy. Many others are larger and grander.

The monks are far better than I am at disciplines such as meditation or contemplation, which require a method of thought, reading, or chanting. However, I've just thought of a type of contemplation that might move our lesson ahead. In the classroom, the red drapes are drawn but bright afternoon sunshine leaks in. The air shimmers with premonsoon heat. A fly buzzes lazily around the room. I break off from my science narrative and announce that we should all be calm and close our eyes.

Here's the contemplative journey I take them on:

Concentrate on your breath and the stillness of your limbs. Listen to the overlapping waves of sound from the cicadas outside and the hum of the generator on the roof. Feel air flowing over your head, pushed by the fan blades overhead. Relax. Be restful but alert. Focus completely on your breathing.

Now shift your perspective. Rise out of your body and lock onto the fly that's moving around the room. Place your entire attention into that tiny creature. You see what it sees. You are moving around the room in figure eights, and looking down, you see shaven heads and robes, mats and a dusty wooden floor. You see a gap in the curtains and fly toward it and through the open window and up into the hot air. You can see monks walking between the buildings and dogs lying listlessly in the shadows.

Suddenly you are engulfed in darkness and feel yourself moving very rapidly. You let your attention swell outward. You are a swift that has swallowed the fly. Swooping and veering, you are looking down on the monastery. You can see the brightly painted roof of the temple and the patchwork of fields and rice paddies beyond. The neighboring village comes into view and the tangle of buses and rickshaws that snarl its streets. You feel a pull upward. Tugged by a buoyant bubble of air, you corkscrew higher and higher, and the fields shrink until they are a fine mosaic of yellows and browns and greens. With a conscious effort, you part company with your avian host and continue to rise with the thermal. A smooth elevator carries you upward.

It becomes cold. You notice the sensation and set it aside. The air is thinner, and steadily, almost imperceptibly, the sky around you gets darker. Below, the Himalayan range comes into view, a crumpled white sheet of enormous proportions. You can see across the Tibetan plateau and over Asia. As the sky becomes black, you keep looking downward. At the edge of your vision, you see a slightly arcing line, the boundary between the biosphere and the black of night. As you continue to rise, the limb curves into a circle and you see the whole planet clothed in its slender sheath of atmosphere.

Moving is effortless. You are unaware of your body and have complete control of your direction and speed of motion. Continuing, you see the Moon and then the inner planets. When it first swims into view, the Sun almost hurts your eyes, but it recedes quickly and becomes a bright dot among many points of light, looking like a moonlit field dense with daisies and buttercups. The solar system is now lost and indistinguishable in the star field and you suppress a moment of panic. But you are confident of your mastery of space and you continue upward, until the majestic spiral arms of the Milky Way are visible and then the whole galaxy is laid below you like white ribbon on black velvet. You pause to take in the scene, suspended effortlessly in the void.

Scientific Visualization

Figure 2.5. To gain an intuitive sense of scales of time and space beyond everyday experience, visualization is a powerful technique. Guided visualization can bring the entire universe down to Earth.

Then I reel them in and bring them home. The class is eager to share experiences felt during the visualization. When the group is engaged and speaking openly, I always find one or two monks who cause me to reflect on their lives here in India and their futures. Geshe Nyima, who played such an active role as our Sun today, is someone I know pretty well, and he shares that he most enjoyed seeing the monastery from above. Next to speak is little Dawa Dorjee (who played our Moon). In conversation, he'll impress me with his confidence in the role of monks in the modern world. "Tibetan culture is the lamp that was earned by our forefathers at the expense of their sweat and blood, and has been passed on to us," he will tell me. "We should not only preserve this golden lamp but pass it on to the next generation. No matter how hard the outside wind, we should protect the lamp while preserving its brightness." For now, Dawa says that during the visualization he was particularly struck by the profusion of stars in the Milky Way.

Another monk speaks, and he also fascinates me. Sonam Wangchuck is a baby-faced monk from the south of India who rarely says anything in class. I pay close attention. The journey was enjoyable, he explains, "But coming back toward Earth I could see India but I could also see my home in Tibet, so I slowed down until I was hanging in space." His voice sinks to a whisper. "I didn't want to come back here."

Soon, I'll have time to meet Sonam and hear his inside story about a Buddhist monk's life. "I think monks should be viewed from two vantage points," he tells me, continuing,

> One is based on someone who just believes in this present life and the other from someone who believes there is a next life. From the first perspective, a monk's life is boring and there are a lot of restrictions such as not getting married and not drinking alcohol. I don't think any reasonable discussion would convince them that a monk's life is worth living since we got rid of all the worldly pleasures. And it wouldn't make any sense to describe the benefit of following a path to enlightenment by progressing over many lives. But those who follow our spiritual tradition can understand the benefits

> of being a monk and following the monastic code; they are much happier and more content in their daily lives.

Conversations such as these help me to understand the complexity of the monastic life and the individual traits of every monk. It's all part of overcoming misconceptions—even stereotypes—that naturally arise between groups that don't know much about each other, especially at what we are calling the "gut level" in some of our science discussions.

Sometimes preconceptions are confirmed. I arrived believing that the monastic life was hard and all the evidence confirms this. Sixteen-hour days begin with chanting at 5 a.m. and end with studying and debating after dinner. Their diet is chapatis or rice for breakfast, rice with vegetables for lunch, and more rice and dal for dinner. Meat and fruit are rare. The food is healthy but rudimentary. It's not surprising the monks sometimes get run down or sick. But what comes as news to me is the hardships from unexpected places; there are serious conflicts among "peace-loving" monks and their fellow Tibetans, and there probably always have been.

One day Thupten B stuns me with the story of his sister, a contender for the best student in the school. She was poisoned and killed by a rival, another girl in her class. The police never investigated. Perhaps I'm naive, but it's hard to imagine jealousy and murder in the majestic setting of the Himalayas. I'm telling this story to Bryce when he shocks me again by saying that the founder of the college hosting the workshop was murdered in Dharamsala along with two of his students in 1997. Interpol is still searching for the killers, who are thought to belong to a Buddhist sect that disagrees with the Dalai Lama and his tradition over the status of a deity called Dorje Shugden. I never imagined that Buddhism could harbor such bitter divisions.

Another "stereotype" is probably pretty accurate, and something Westerners from a university culture must deal with in a Tibetan setting. The traditional culture is very patriarchal. This comes out one day in the kitchen, where a woman is assisting the Tibetan cook for the workshop. She's arguing with him in a very intimate way, so I ask our organizer, Bryce, if they're married. Bryce smiles. We're embedded in the ordinary monk's world of chastity but obviously not all Tibetans are required to be celibate. Far from it. The woman is married to the cook but she's also married to a tall, gaunt man who works in the garden. His nickname is "Uncle." Polyandry is an enduring part of Tibetan culture, though it's fairly unusual in India. If a man has two sons, it's an advantage to split the land between them and have it stay in the family when they marry the same woman. Who am I to judge? The woman in the kitchen seems to handle two husbands with ease.

The only woman among the teachers, Gail Burd, adds another perspective on the life of the all-male fraternity of monks. The first observation is gender-neutral. As the biology instructor, she's learned that news of DNA and the genomic revolution hasn't reached most Tibetan Buddhists. In teaching them developmental biology, she's also uncovered some surprising misconceptions. Some of the monks attribute sentience to plants and trees. Then we laugh as she muses on the lack of testosterone in a room full of young men. I ask Bryce why there aren't women in the program. He looks pained. We planned on having nuns, he says, but their exams conflicted with the timing of the workshop. I'm at least glad to hear Buddhism accepts women into the fold. But it's been a bumpy road.

This becomes clear when Paul and I drop in on one of the weekly classes on advanced Tibetan philosophy held at the Library in Dharamsala. The room is packed. Two-thirds are Westerners, and most of those are American. Some have the look of moth-eaten hippies. Others are wearing expensive yoga gear, bearing the Prana label, durable and fashionable (though not quite Prada!). They're not casual practitioners. For the fifteen-minute prayer that precedes class they chant in Tibetan, without notes.

The lecture is given by a wizened, Yoda-like monk who's less than five feet tall but who peers down at us from on top of an ornate shrine. He's translated by a woman with an English accent. Her gray hair is tied back and she has piercing blue eyes. He reads text from an eighth-century interpreter of Buddhist thought. It's okay to be rich and powerful and handsome, he cackles delightedly. You can draw people to you and do good things in your life. It's also best to be a man. More than half of the people in the room are women and they laugh uneasily. The translator also laughs, then vigorously conveys her disagreement under her breath.

According to other traditional texts, Gautama Buddha was asked by his aunt and stepmother, Pajapati, if she could be ordained as a nun. Although Buddha saw no reason that women couldn't be enlightened and enter nirvana, he resisted. But Pajapati was persistent. She gathered five hundred women. They cut off their hair and dressed in patched-together robes and followed Buddha on foot for hundreds of miles. Buddha's cousin helped Pajapati make her case, and he finally relented. Given the status of women twenty-five hundred years ago, his decision was probably an indication of enlightenment. But he gave an ominous warning that ordaining women would cause his teachings to last half as long, and even though women could be ordained they had to follow 350 rules instead of only 250 for men.

The tradition of fully ordained nuns, or Gelongma, died out in Tibet and India for centuries. Then, in 2007 an international congress of Buddhist scholars recommended their reinstatement. Its conclusions were supported by the Dalai Lama. I'm glad to hear it, since how can we experience the sky without the half that's held up by women?

Another stereotype is worth mentioning, since I found myself stumbling over it a few times as I reflected on how to teach Buddhist monks. Not only had I held the view that they were unequivocally peace-loving, but I'd observed that their educational methods were unidirectional: the teacher speaks and the monks take it in. Period. That was until I happened to see Buddhist monks engaged in formal debate.

The night after arriving at the College for Higher Tibetan Studies, I had bad insomnia. So I got up at dawn and walked through the ramshackle campus. I heard the sounds before I saw their source: percussive claps, shouts, and peals of laughter. On the far side of the central courtyard was an area of grass and acacia trees. Through the dim, hazy light I saw pairs of monks, one of each pair sitting on the grass and the other standing. I watched, transfixed. The routine was stylized yet free-flowing like a form of dance. Despite the intensity and the noise, these debates held no aggression. Yet there were clearly winners and losers. Some monks acted triumphant and exhilarated while others looked pained and hung their heads.

Buddhists consider that all suffering is related to a failure of wisdom, even the suffering of death. Philosophical debate is a tool to see the nature of things and achieve wisdom. It's a core part of the training of all Buddhist monks. In India, it's taken seriously enough that losers in a debate must accept the view of their opponents. Debate is fluid and dynamic, with choreographed rhythm and grace.

It goes like this.

The Challenger approaches the Defender respectfully with a quandary. He stands while the Defender sits. The Challenger poses his question with his right hand raised above shoulder level and his left hand stretched forward with palm upward. At the end of his statement, the Challenger claps his hands loudly and stomps his left foot. He then quickly draws back his right hand with the palm upward and leaves his left hand forward with the palm downward. The left hand represents wisdom and the right hand represents method, particularly the practice of compassion. Bringing the two together is the union of wisdom and method. At the moment of the clap, the foot comes down hard to slam the door on rebirth at the lower levels. Holding out the left arm keeps shut the door to all rebirth. Drawing back and raising the right hand symbolizes the will to raise all sentient beings out of cyclic existence and grant them the omniscience of Buddhahood.

Monks practice debate for two hours in the morning and two hours in the evening. They practice year-round, indoors and out, through the depth of winter, and in darkness lit by candles. Excessive prior preparation is frowned upon; debaters should live by their wits and argue in the moment. It occurs to me that debates are probably loud and active for a purpose, since monks are celibate and aren't allowed to play organized sports. They literally live in cells. Debating is an outlet.

Gelek gives me and the class a great example of the kind of topic that's debated. "We commonly debate the nature of a cup of water," he says. "The gods may see it as ambrosia, while human beings see it as water, and a hungry ghost sees it as full of blood and pus. Such discussions often take place in our debates, and it was very difficult for me to comprehend or believe in these statements." He continues,

"The debate question arises because the visual perceptions of the god, the human, and the hungry ghost are all incontrovertible [to their subject viewpoints], and yet what is contained in that single cup has not objectively changed either—so what is the reality?"

Harvesting Old Light

That night at the debate, I proposed a statement where the monks could take sides as Challengers and Defenders. It was the assertion, "We can understand what is happening *now* in the universe." The arguments, slapping hands, and stamping feet wheeled into the night, but it was only the start of our discussion. The proposition is related to the speed of light, and after today the monks will have more facts as fuel for their debate.

We return to our scale-model of the universe. Before class, I walk around the campus physically positioning the scale model objects of the inner solar system. I'll be holding the toy Earth in one hand and the small gray marble in the other to represent the Moon. The red marble of Mars is visible on a ledge at the far end of the building, and to position the Sun, I've had to go down the dirt road a ways. There, I've propped a large yellow paper Sun against an acacia tree, which will be just visible from our classroom window.

Once we have all these bodies in view, I tell the monks that if light

Lookback Time

If the speed of light were infinite, light from everywhere in the universe would reach us at exactly the same time and we would see the entire universe as it is now.

But it is not, so we see distant regions as they were in the past.

Distant Light = Old Light

Figure 2.6. In cosmology, telescopes act as time machines because the farther out we look in space, the further back we look in time. Due to the finite speed of light, the history of the universe is visible.

went infinitely fast, we'd see everything in the universe as it is right now—not just the Moon I'm holding here, and Mars there on that ledge, but also the Sun down the road. And the stars that in the model are tens of thousands of miles away, we'd also see them instantaneously. This makes sense to them, and they're nodding. Then I explain the dilemma. If light traveled infinitely fast, we'd be overwhelmed by all the light emanating from trillions of stars and billions of galaxies. It would hit our eyes all at once. Total information overload. At this, the monks don't look worried but rather puzzled.

I tell them that the answer was given by Albert Einstein, the physicist. He postulated that there's a constant and finite speed of light. In our scale model, its speed is greatly reduced, and it can't travel instantaneously. Instead, it strolls. So if a flare erupted on the Sun and a jet of magnetized plasma looped out from its surface, the light showing it would take eight minutes to walk to us. And if the planet Neptune suffered a major impact that decimated its surface we wouldn't learn about it for five hours. I lean forward conspiratorially. Or, if an evil monk empire from the other side of the galaxy destroyed our Sun—I point toward our paper model down the road—its light and life-giving heat wouldn't go out for eight minutes. There's laughter around the room. I assume they're laughing at the very idea of an evil monk empire.

Then we consider light from another galaxy. I hand around a picture of M31, the great Andromeda spiral. Dawa chimes in that it looks like the Milky Way did at the end of our visualization journey. Its light takes 2.5 million years to reach us. Strange as this may sound, it makes no sense to ask what the Sun is doing *now* or what the M31 galaxy looks like *now*. We might wonder, but we can't know. We have to be patient, and that information about the *now* will reach us in due course. By which time something else will have happened there. When it comes to the universe, we're stuck reading yesterday's news.

I ask them to consider how this strange time lapse in viewing the universe relates to human existence. Today we see the light of the M31 galaxy as it was when our ancient ancestors *Homo habilis* lived in a small corner of Africa. The light of M31 was still streaking across space while humans evolved a larger brain, and thus became *Homo erectus*, leaving Africa for the first time. The light reached the edge of the Milky Way disk as anatomically modern humans fanned out over Europe and Asia and were settling Tibet and North America. The light angled in over the Perseus spiral arm as the first cities were being built in Mesopotamia. It whipped by the luminous blue star Almach as New Amsterdam passed into British hands and was named New York City. It grazed close to the Sun-like star Lambda Andromeda as Edwin Hubble, in the 1920s, first measured the distance to its point of origin. Eighty years later the light from M31 finishes its long journey by arriving at the Earth's surface.

So, I tell the monks, if you know where to find M31 in the northern sky, when you look at it now, the light that enters your eye now began its journey long, long ago. Distant light is old light.

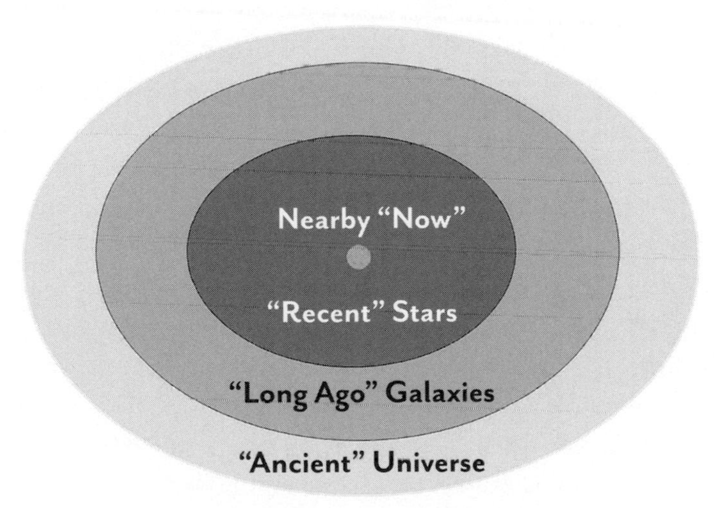

Figure 2.7. We see the Sun as it was eight minutes ago, nearby stars as they were dozens of years ago, nearby galaxies as they were millions of years ago, and the distant universe as it was billions of years ago.

Looking around I see perplexed expressions. The concept of "look-back" time is counterintuitive. We put it into practice. What if four monks were standing on planets remote from each other? Earth-like planets are known to be fairly abundant around other stars, so what we'll do next is plausible. Dawa stands next to me to act as the Earth. Thupten stands nearby, but he's ten light-years away in a similar Sun-Earth system. I send Jigme (who demonstrated quantum uncertainty for us on a previous occasion) across the room to be another Earth-like planet forty light-years away. Then I direct Gelek to the back of the room, eighty light-years away. He's our fourth and final analog of the Earth.

But now we need photons—particles of light. Four more monks volunteer. They're information carriers, traveling at light speed, which in this classroom experiment is a slow shuffle. For this and for other purposes I've cut sheets of fluorescent yellow paper into jagged star shapes. Each monk who plays a photon will carry one.

Our time-lapse problem will be this: the three monks on distant planets want to celebrate their most recent birthdays with their friend Dawa on Earth. How can this mutual celebration be communicated, given the distance between the planets and the finite speed of light? The three birthday monks draw pictures of themselves on the yellow paper. Then they label them with their names and ages. These photons are birthday cards that will be mailed at the speed of light. We start the experiment.

The three monks who are playing photons take the yellow birthday cards in hand and, as if light beams, shuffle toward Dawa at the same speed. After what would be ten years of light travel I tell everyone to stop. Thupten's photon, who's carrying a card for Thupten's thirty-fourth birthday, has reached Dawa, having traveled just ten light-years. But there's something wrong with Thupten's birthday card and I ask Dawa what it is. He says that by now, Thupten is forty-four years old. "Exactly," I say. Thupten aged ten years while the birthday information traveled.

We resume, and the other photon monks shuffle their way to Dawa, and by now the pattern is clear. Jigme's birthday card, celebrating his twenty-ninth, gets to Dawa when Jigme is sixty-nine. Gelek's birthday card has to travel eighty light-years, so Dawa must draw the obvious conclusion. How was Gelek as his card arrived? "I'm afraid he's prob-

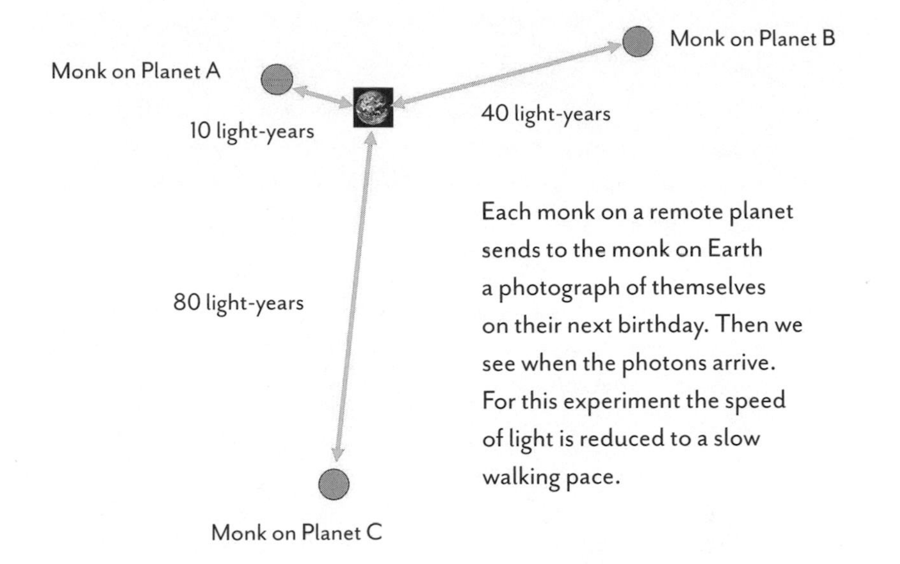

Figure 2.8. No information can travel faster than light. In acting out the way signals travel to planets that are dozens of light-years away, it makes no sense to ask what is happening on a distant star "now."

ably dead," Dawa says sadly. Across the room, Gelek looks crestfallen but I give him this scant consolation: Dawa has been waiting the same eighty light-years, so he's probably dead, too.

In the far future, this problem of human life spans will be the obstacle to having interstellar pen pals. A letter from even the closest Earthlike planet will take at least ten light-years to arrive. The playful monks absorb this outcome equably; they seem to have no fear of or reverence for death. I try to picture the absurdity of interstellar chat for the class. Here's Lobsang Choephel on one planet writing to Thupten: "Greetings. What are you doing?"

Thupten replies, "Visiting a friend in Bangalore."

Lobsang writes back, "Tell me about the trip."

Thupten replies, "Actually, I've forgotten. It was twenty years ago."

In the larger realms of the universe, light travel times are enormous. We see the center of our galaxy as it was thirty thousand years ago. We see the galaxy M31 as it was 2.5 million years ago. And this is all in our own galactic backyard. Light from the nearest big cluster of galaxies in the Virgo constellation has been traveling for 50 million years, since the time the first birds evolved on Earth. The huge Coma cluster of galaxies is seen as it was 150 million years ago, when Earth was ruled by dinosaurs.

All of these look-back times are small fractions of the 13.8-billion-year age of the universe. But by using large telescopes we can easily gather light from galaxies 11 or 12 billion light-years away. We see these galaxies as they were long before the Earth formed, when they and the whole universe were young. The beauty of astronomy is the use of telescopes as time machines.

Contents of the Universe

Without Tenzin Sonam at my side, I'd have great difficulty conducting our daily classes. At the end of my first week of teaching, I realize the profound importance of clear translation. Each day we plow into more complex concepts, requiring more scientific terms, and pitching us into ever more convoluted discussions. What would I do without Tenzin?

Tenzin has a profound story that goes with his exceptional linguistic skills. His parents fled Tibet in 1961 and settled in Dharamsala. After getting a degree in engineering, Tenzin became one of the first staff members in the Library of Tibetan Works and Archives to work on science translation. Since he has no background in cosmology, we've talked extensively about how to handle jargon and subtle concepts. Beyond that, the relationship is based on trust. I speak in short bursts, several sentences at a time. He translates without hesitation. I watch their faces to see if the monks' expressions are clear or cloudy. Questions and answers are handled in the same way. By closing the loop with discussion, I can ensure they've truly understood.

Nevertheless, teaching from one language to another may be the most inefficient form of education known. I'm chafing at the fact that it takes twice as long to cover the material this way. It's also frustrating not be able to keep up with the discussion among the monks. They're inside the candy store, feasting on the morsels of knowledge and living it up with spontaneous and raucous camaraderie. I'm the kid with his nose pressed against the candy store window, wishing I could get inside.

Meanwhile, Tenzin sews together our verbal exchanges as seamlessly as possible. We return to our topic of the material content of the universe. "How do we estimate the content?" I ask rhetorically. The first thing we should know is the limitation of the human eye. That's why we now rely on electronic detectors and telescopes. They pick up electromagnetic radiation we can't see. I ask the monks why telescopes are better than eyes.

"Bigger!" says one monk.

Rather than reply immediately, I hand out rulers. In pairs, the monks carefully measure each other's pupils, at a slight distance for safety. The pupil of the human eye is an aperture. It's black because light goes in but doesn't come out (being absorbed by pigments in the retina). The average monk pupil is four millimeters across, and we adjust that to eight millimeters since a pupil opens wider in the dark. A telescope is also an aperture, and a medium-sized telescope is two meters in diameter. So the telescope has about sixty thousand times more area than the pupil, a large gain in light-gathering power. In modern telescopes, the light hits electronic detectors (and not human eyes) that convert virtually every photon arriving through the telescope aperture into digital data. For all the wonders of the human eye, its retina only registers a small percentage of the arriving light.

What about a camera, I ask the monks? What advantage does it have over the eye? They get this one quickly: cameras have variable exposure times. With a camera behind a telescope, astronomers can open the shutter for fifteen or twenty minutes or more, and collect much more light from very faint objects. To see how much more, we pause to make thaumatropes. This toy, popular in Victorian times, is spun until two images visually merge together.

Making a thaumatrope requires a circular piece of white cardboard, which I began handing out to the monks. With little explanation I ask them to draw a picture on both sides, requiring only that the two pictures are centered on the same point and that they complement each other. I give an example: a flower on one side, a flower pot on the other.

The monks split into groups and the designing begins. They do pairings of a face and its features, a bird and a cage, and the base of a temple and a tiered roof. Most elegant of all is a Buddha and a surrounding altar. When the cards are complete, each one is attached to the end of a thick straw so that they look like large lollipops.

I ask them to spin the straw between the palms of their hands and watch what happens to the two pictures as the speed is varied. The room is alive with chatter and laughter. They realize that when the card spins above a certain rate, the images on either side of the card merge into one—it's called persistence of vision. That rate matches the time the eye gathers light before "resetting" and forming a new image, which is like the exposure time for a camera. It's not easy for the monks to measure the spin rate at which the merger takes place, but it's a good creative struggle. Eventually, a few groups manage a crude method of measurement. They measure how far the straw rolls per second along their palm and divide it by the straw circumference to get a spin rate. The average answer is ten times per second. So the brain registers what's falling on the retina every tenth of a second. To properly compare the power of light gathering in the eye with electronic detectors in a telescope is complicated, but I assure them that the two aren't even close; the best telescopes have the power to gather 10 billion more times light than the eye.

Don't be discouraged, I tell them, for today we are going to use our eyes to the best of their powers. We're going to look carefully at the best image of the sky ever made by a powerful modern telescope. The Hubble Ultra Deep Field image was produced by the Hubble Space Telescope while it was orbiting in space. Each monk gets a single high-quality color print of the Deep Field image.

In making this image, the Hubble Space Telescope wasn't pointed at any known galaxy or cluster. It was designed to peer at a boring part of the sky so that the census of galaxies would be representative of the whole universe. It stared for several weeks at a tiny patch of night sky.

The image is disorienting, crowded, and even fuzzy at first. As the monks inspect the image, I tell them that each of the fuzzy patches of light, down to the faintest smudge on the paper, is a vast stellar system billions of light-years away. The patches are nestled cheek by jowl in the picture but that's an illusion. The space between them is enormous. They're close together because the image collapses three dimensions into two. It's like looking at a sparsely planted grove of pine trees from the side; viewed from above they have large spaces between them, but seen from the side they'd appear crowded together. The Ultra Deep Field photo is a long and skinny pencil beam, a core sample of the universe stretching back 90 percent of the way to the big bang.

I finally tell them their task. They're to count every one of the galaxies in the photo. The monks are brave explorers of space and time; they don't flinch. Each faint splotch is a distant stellar system snared by the Hubble Space Telescope, like dewdrops suspended in a dark cobweb. Many are so small they register as a few pixels of pale white on the black background. Others are big enough to show spiral or elliptical shapes and subtle colors. Looking out at my universe of monks, I

Hubble Deep Field

Many systems of stars like the Milky Way.

Figure 2.9. The Hubble Space Telescope stared for several weeks to produce the deepest picture of the sky ever made. Every image in the tiny patch of sky is a distant galaxy, and most are billions of light-years away.

witness a sea of bobbing shaved heads. They've organized into groups of three or four according to home monasteries. They hunch over their prints and count with exceeding care. The room is quiet except for a low communal murmur of counting that ebbs and flows. I close my eyes and it sounds like chanting.

Knowing the patience of a trained monk, I believe some of them could actually succeed at counting each galaxy, even though there are several thousand. However, Thupten B has found a better method. To keep track in his group, he divides the Deep Field image into quadrants. Each monk counts in a different sector. I interrupt and point this out to the other groups, explaining that it provides a sampling that is perfectly adequate for a final estimation. The rest of the groups switch methods and a period of intense quiet follows. Then they call in the numbers. Each of the six groups provides a slightly different estimate of the total number of galaxies in the image. The average is twenty-eight hundred.

That's how many galaxies the Hubble Space Telescope saw in its very narrow line of sight. In fact the idea of its field of view being pencil thin is an overstatement. In terms of the entire sky, the Hubble Ultra Deep Field image represents the size of the head of a pin held at arm's length. This tight view illustrates the basic trade-off of astronomical optics: for greater depth, you must narrow the field of view, and vice versa. Large and powerful telescopes see far by narrowing their field to a tiny patch of sky. For example, the most sensitive camera on the Hubble Space Telescope has a field size of "three minutes of arc," just a tenth of the Moon's diameter in the sky.

By relying on such a small patch of the sky, astronomers make another important assumption. They assume that the universe is uniform in all directions. This assumption is called the "cosmological principle." It allows us to get a complete census by sampling a tiny patch of galaxies and multiply that by the number of patches that can fit on the entire sky. The monks counted twenty-eight hundred galaxies in the Hubble Ultra Deep Field patch in the sky, so we now need to know how many such patches make up the entire sky, north and south hemispheres, which represents a full view of the universe. For this calculation, we take the classroom wall as the entire sky. According to my earlier calculations, the Hubble Ultra Deep Field patch is on that scale the size of the head of a pin.

I can't help waxing lyrical when I ask the class, "How many galaxies can fit on the head of a pin?" Then I quickly explain my point, gesturing to the wall. The number of times a pinhead can fit on the wall is about equal to the number of times the Ultra Deep Field image fits across the entire sky. I can see it in their eyes. They know I'm going to ask them to start measuring pins on the wall—which is true.

They measure the wall in square centimeters. Next, they find the number of pinheads per square centimeter. They conclude that it will take 3.5 million of them to fill the wall. In other words, the Hubble Ultra Deep Field image represents less than three millionths of the total sky. Multiply that by our 2,800 visible galaxies and you have 10 billion. This is the largest number we've produced in class; ten times more than the number of sand grains in the bucket. But even the Hubble Ultra Deep Field doesn't detect everything. It includes large and small galaxies in the closest sectors of the universe, but for the distant sectors it only sees the largest and brightest galaxies, so the census is an underestimate. Typically, the census doesn't reach to the limit of distant light; it undercounts by a factor of two in distance, which is a factor of eight in volume. So astronomers increase the normal see-and-count estimate eightfold. That makes 80 billion galaxies if we use the monk's tabulations. This is 20 percent lower than astronomers' most careful calculation of 100 billion. But it's a darn good estimate. The monks' census is an unqualified success.

Before they feel too satisfied—which they clearly do at their patient accomplishment—I tell them that the numbers still get bigger if we consider the stellar content of the universe. For this, we return to the familiar metaphor of grains of sand. If each grain stands for a star it would take four hundred buckets—enough to fill the classroom with sand a foot deep—to represent the stars in our galaxy. But there are about 100 billion galaxies in the universe, so we have to imagine 100 billion classrooms a foot deep in sand or a billion times that area a hundred feet deep in sand. We close our eyes and do a brief visualization: think of standing in the classroom filled with sand. Wiggle your toes and feel the stars underfoot. Each one is a tiny system of a stars and whirling planets. Now let the walls of the classroom melt away. Extend your sandbox into a vast playground three hundred miles on a side. That's the stellar universe. There are more stars in the universe than there are grains of sand on all the Earth's beaches.

Each galaxy on average contains 100 billion stars, so the stellar census of the universe is 10^{22}. With this crescendo, I bring in the question of alien life. Isn't it likely, with many planets orbiting the 10,000 billion billion stars out there, that the Earth is not the only one that hosts biology? This concept is not preposterous to them, for Buddhist training has always taught the idea of countless world-systems in space. The nearest billion world-systems with sentient life are said to be under the influence of the Buddha. While bold, the idea even makes scientific sense. I can easily imagine an advanced civilization that can control or influence 1 percent of the galaxy. Buddhism assigns no limit to the world-systems; they're potentially infinite in number.

The monks are pondering their Hubble images. I ask them, "How likely is it that someone or some *thing* is staring back at you from one

GRAINS OF SAND

"To see the world in a grain of sand
And heaven in a wild flower."
—William Blake

Figure 2.10. Sand grains are a metaphor and analogy to understand the vastness of the cosmos. Each sand grain is a distinct world when viewed through a microscope, and there are as many grains on Earth as there are planets in the universe.

or more of those fuzzy patches?" Then I put aside the restraint we've showed so far in our numbers. Time for full throttle. I tell them that those stars contain 10^{80} atoms, and that for each atom, there exist an additional 100 million microwave photons as leftover heat from the big bang explosion that began the universe. We can't count the 10^{88} photons, even in principle, because unlike grains of sand or galaxies, they're identical. They slip through our fingers at the speed of light. We've finally hit the red zone in our big numbers. And we haven't had time to talk about dark matter, which contributes an added 10^{81} particles, called "dark" because they don't interact with light but only make their presence known because they exert gravity.

Suffice it to say that with the big bang photons, we've reached the largest number that can be calculated in cosmology (although, granted, it's dwarfed by the Buddha's staggering 10^{421}). We decelerate. We take a well-deserved rest.

○○○○○

My recreational running in India has brought nothing but odd stares. Near the end of the first week I go out for the first time. Back home I'm accustomed to the low humidity of the desert and canyons near Tucson. India offers just the opposite, and I struggle through warm, moist air, up the winding hills toward Dharamsala. My reward will be the great views at the top. One intervening village is aptly named Gaggal. It's just a gaggle of simple houses strung out along a dusty street.

The shops are single rooms with one wall missing, open to the street, and lit by a naked bulb. They all seem to have exactly the same selection of dry spicy snacks, toiletries, and sodas. Children recite lessons in singsong in a one-room schoolhouse with dirt floors and no glass on the windows. A roadside temple has crumbled statues covered with moss and ferns. Gaunt men squat idly at the roadside on their haunches. Women are doing most of the heavy labor—washing clothes, tending cows and goats, and staggering under large plastic flagons of water and rice.

Their eyes turn to follow me as I pass. Exercise is an alien concept. It's hot and humid. Who in their right mind would run when living takes most of your energy? I wave and smile. My message: I'm crazy but I'm harmless.

I'm also thinking about what has happened so far. I'm bruised on the inside. My divorce, however amicably arranged, continues to echo as a wrenching experience. After lying low for a couple of months, I started dating and found it a disconcerting and sometimes surreal experience. Maybe I was born to be a monk, either in this life or the next, or the one after that. For the time being, it suits me quite well to be "cloistered" in this remote setting by the Himalayas.

Separation from my normal work life is leading me to question things I rarely question. I'm in the upper reaches of my profession, with the power and responsibility that entails. My colleagues are very smart. I get to work with excellent students. Large telescopes are on my doorstep. But I'm also hemmed in by "the system," in the words my son Paul often uses. I'm often enmeshed with administrative work and institutional bureaucracy. There's the relentless chase for research funding. It wears me down. In the drive for professional success I've lost sight of some of the pure elements that got me into astronomy.

I think of Socrates. He used to walk around Athens teaching his students, a peripatetic philosopher. So now I'm running with Socrates up a hill in India. Why not? He was nearly a contemporary of the Buddha, and he nailed it on the head when he said, "The unexamined life is not worth living." How will my life look when I see it through fresh eyes? The monks are rekindling my sense of playful curiosity.

Being with Paul in these exotic new circumstances, and going through them together, has been an opportunity we probably couldn't have found in Arizona. We're renewing and deepening our bond. I like the fact that the locals know with a glance that we're related. We're both settling in as resident aliens. We dutifully take our Malarone to fend off malaria. We've arrived with armor against polio, hepatitis A and B, typhoid, tetanus, and diphtheria. Our internal systems are still adjusting. We've beat back the worst; my digestive system is holding the line, but Paul still has the normal runs and nausea of an outsider eating local food. I'm afraid he's getting weaker, not stronger. Luckily he's brought a pile of books. Three are about the Dalai Lama. So he has plenty of diversion for a while.

On an evening when Paul's feeling better, we go to the temple where the monks are celebrating Sagdawa, a major lunar holiday. It commemorates the birth and death of the Buddha and tax day, too. The monks are energized since they've been preparing for days. Chanting commences, and everyone's given a package of food and drink containing a juice bag, a mango, some Bombay mix, and a stupa made of ground barley and dried fruit, which looks slightly suspicious (Paul calls it a yak turd). I remind him that this is a "multiplier" day, when good and bad deeds are magnified, so we can't afford any bad karma from spurning the offerings.

When the prayers end, the monks continue the celebration by holding a debate. At the front of the room the geshes sit serenely and confidently behind a dias. Taking turns, the younger monks approach in waves, their bodies thrusting and parrying in sync with their arguments. Even without translation, it's high theater and compelling. One of the monks named Kalsang Gyaltsen catches my attention. He told me his story recently.

Rising from rural poverty in Kham Province of eastern Tibet, he bore the brunt of responsibility for his family; his siblings and grandparents couldn't work, and his mother was seriously ill. They lived in a Tibetan collective, where land, livestock, and furniture were gathered and shared. Such commodities, and basic food like cheese and butter, were redistributed according to a family's ability to do work. Then his life changed suddenly. A movement had arisen among Tibetans to rebuild their destroyed monasteries. The tug of karma, Kalsang says, led him to become a monk. Initially his work was heavy construction. There's no equipment; everything must be done by hand. Those were full days indeed. He got up before dawn to memorize religious texts, worked from sunrise to sunset, then studied until ten at night. He became a jack of all trades, trying his hand at wall-building, carpentry, religious dance, and making sand mandalas. With a barrel chest and a deep voice, Kalsang Gyaltsen became a specialist in chanting, the kind of throat singing unique to the Asian plateau.

Now during the Sagdawa celebration, he's on the debate floor, silent but letting his gestures express his argument. The monks are by turn combative, exultant, tremulous, and preening. There's plenty of piss and vinegar on display. In the face of a forceful thrust by his opponent, Kalsang stands his ground. Then suddenly everything stops. The entire Sagdawa celebration has reached a peroration, and rather than wind down it's simply over. There's no closing ceremony. Everyone just leaves. It's the Westerners who linger awkwardly and outstay their welcome. Tibetans, I think to myself, do endings well. Socrates would be impressed.

CHAPTER 3

Beginnings and Endings

IT'S ONLY TEN MILES up the winding road to Dharamsala, but looking through the windows of a dilapidated taxi, time seems to have slowed down. As we drive through small villages, men are sitting patiently tending little street shops. Women walk miles carrying baskets of lentils or plastic flagons of water on their heads, and in the fields families maneuver twin oxen that plow a single furrow. Old men squat listlessly by the roadside. A young child in rags draws geometric patterns in the dirt with a stick.

Only automobile driving in this Tibetan enclave is laced with urgency. Our driver is slow and careful, but the other taxis seem hell-bent. They careen around hairpin bends and rush through villages where crowds of people miraculously divide to let them pass.

Time has slowed down for me. How can it be so different? At home it rushes by, but now it oozes, almost dreamlike. One obvious reason is the absence of cell phones and the Internet, which by nature are insistent, nagging robots. I'm surprised to find I have more patience. Part of this is spurred by the monks. They're embedded in the rhythm and ritual of monastery life. They never seem rushed; they never seem to look ahead to the next thing. Entering my second week in India, acceptance and tolerance are seeping into me, like strange new friends. I'm more fully in the moment than any time I can recall.

As the taxi approaches Dharamsala, we see glimpses of the Himalayas, mountains that reflect the epitome of time. To the north, they're snow-cloaked even in summer, rising up to sixteen thousand feet. The proximate peaks are majestic but are merely foot soldiers of the Himalayan range, which spans a thousand miles and rises nearly twice as high again with the crowning point of Mount Everest. Everest is the name given by Europeans—Earth's tallest mountain is Chomolungma in Tibetan, which means the Goddess Mother of the World.

Dharamsala is actually two towns. Today we're visiting the lower town, which is more dusty and chaotic. It has no tourists and few Tibetans. It's the best place to buy our provisions. Leaving the taxi with instructions to pick us up in an hour, Paul and I roam the winding main street with its cacophony of car horns and merchants shouting, and its tangle of naked electric wires dangling just overhead. We look for toilet paper and soap and mosquito coils. We haggle because it's expected. But prices are so cheap it feels like more of a ritual than a necessity. A woman wearing a gold and green sari sits on the ground on a blanket surrounded by fruit. She flashes us an impossibly white smile so we buy some of her mangoes.

The next stop is McLeod Ganj, named after a lieutenant governor from the colonial era. McLeod Ganj is a tangled warren of streets teeming all summer with pilgrims, backpackers, and unreconstructed hippies. The Tibetan government-in-exile occupies a site a mile from the town center. Tsuglagkhang, the major temple where the Dalai Lama lives and worships, is a big attraction. Local Indians rub shoulders equably with several thousand Tibetan refugees.

One of the translators is Sangey Tenzon, a cheerful and eager fixer of sorts. He has arranged our taxi travel today with great efficiency. My son is still adapting to the food and climatic change, and we're on a mission to buy some staples that will settle his system. One staple that doubles as comfort food is English biscuits, and not just any, for Paul knows that Hobnobs and Digestives are the best. Sangey shows us around. He's in his mid-twenties, a smart dresser with a round face and mop of shiny black hair. Sangey leads us into tourist shops selling thangkas and carvings, and he always seems to know the owner. He says he can get us a special deal. We demur.

On another excursion, I'll learn more about Sangey's personal history by visiting the nearby Tibetan Children's Village. It's the school he came to as a nine-year-old, when his parents deposited him with a group making the perilous journey across the Himalayas. Life in the children's villages has always been incredibly hard, but recently it has become better, Sangey said, though most buildings are still without heat in winter. The older children create solidarity by playing in the rice fields and forests. But my heart aches when I see the nurseries, lined with orphans sleeping head to toe like sardines, under blankets in large cribs.

After we finish in Dharamsala, loaded down with Her Majesty's biscuits, we return to the campus, and I prepare for another day in class. The topic will be time on the cosmic scale. We arrive just in time for the tea break, where I learn that Buddhism has its own time measurements. Geshe Nyima explains to me the most enigmatic unit of Buddhist time, the great kalpa. It's analogous to an eon and was borrowed from Hindu tradition. It's roughly equivalent to a trillion years. Buddha never talked about the exact length of a great kalpa. Instead he gave an analogy. Imagine a granite mountain a mile high and a mile wide at the base. Once a year a dove flies by it and brushes it lightly with its wing, dislodging a few grains of rock. A great kalpa is how long it would take for the mountain to be completely eroded by the wing of the dove.

The Nature of Time

As the kalpa story suggests, deep time is nearly unfathomable. In our daily experience it's also quite weird. For people without a cesium atomic clock, time will inevitably be subjective. We're back in class, and to explore this strange reality of time, I present the monks with a simple experiment.

First, I show a picture with fifty stars on it and ask them how many are there. That takes a few minutes but everyone nails it; counting is not a problem. Then I ask them to put twin marks on a long piece of paper to define fifty centimeters. Even without a meter stick to help, almost everyone is within 10 percent, so measuring familiar spaces is easy, too. Now I tell them to close their eyes and silently count off fifty seconds, raising their hands when they get there.

Hands go up, and the translators and I read off the times and tabulate them on the whiteboard. The lowest is 29 seconds, the highest is 118 seconds, which is a range of a factor of four, and the dispersion is 50 percent. Why are we so bad at measuring something so familiar? Time's apparent subjectivity—flying by when we're diverted and eking out when we're bored—is just as central to our experience as the fact that time passes at all.

Time has a direction, and that attribute is equally profound and puzzling. Space is omnidirectional; you can go whichever way you want in each of its three dimensions. All other physical properties show no time bias. Objects can rotate one way or the other. Charge can be positive or negative. And from pure energy, you can equally well make matter or antimatter. Time is from a different universe, it seems. It inexorably separates what can be anticipated from what can only be remembered. With no choice in the matter we ride the gossamer-thin membrane of "now." Listen to Argentinean writer Jorge Luis Borges: "Time is a river which sweeps me along, but I am the river; it is a tiger that devours me, but I am the tiger; it is a fire that consumes me, but I am the fire."

In Buddhism, time is neither mental nor physical, but it belongs to a third class of phenomena. Whatever time might be, its experience of being relative and subjective is uncontroversial in Buddhism, whereas Western philosophy has contended with it from every angle. What seems indisputable is that time is like an arrow that goes in only one direction. To spur thinking on the "arrow of time," I show the monks a video that mashes up a series of everyday scenes, set to jaunty, indie pop music.

We see a kid drilling a hole in wood, opening a can of sardines, riding a bike, sweeping a porch, dropping a glass bottle onto cement, stirring powder into a glass of water, blowing soap bubbles, burning a piece of paper, and wrapping a present. We see traffic on a highway, fireworks going off, a dog eating from a bowl, a kettle boiling, and toast popping. After watching the forward motion, I show a version of the video created to play backward. Each event runs in reverse. Some of the scenes look only a little bit strange in reverse time; others look completely ridiculous. The question is why?

I play the video twice more and ask the monks to note which scenes they think can actually occur with time reversed and which could not. This leads to a lively discussion that lasts over half an hour. They all think it may be possible to arrange for cars to drive backward or for somebody to ride a bicycle backward. Rewrapping a gift is fairly normal, too. But wood shavings don't jump back into a hole to leave unblemished wood. Glass shards don't fly off the floor to form a bottle at the edge of a table, and flames don't converge and convert ash into clean white paper. Then there are the ambiguous cases. When raking leaves, for example, the arrow of time seems to operate only weakly. Sonam Choephel and Thabke Lodroe also think a dog eating food is an ambiguous case. They vigorously debate how a dog could regurgitate food, remaking a time-reversed dog dinner.

Now suppose, I say, that an alien unfamiliar with humans sees the reverse video on the galactic feed of YouTube. How would you explain why some scenes are never observed? What's the common denominator in scenes that are time-irreversible, versus scenes that might be seen playing out with time reversed, or could be arranged to do so?

Thupten is first to offer an opinion. The answer seems to be the distinction between macroscopic objects with particular overall motion—a car, a bicycle—and larger numbers of small or microscopic objects with random or complex motion—shards of glass, shavings of wood, smoke, and powder mixing in water. The ambiguous cases are, appropriately, intermediate-sized objects like leaves and dog kibble. This division doesn't amount to a scientific explanation, but it's an important clue. The clue involves order and disorder.

Disorder is familiar in the everyday world; we spend much of our lives combating chaos. In physics, the tug of disorder provides the best framework for understanding the arrow of time. There are many ways for atoms and molecules to be disordered, but relatively few highly ordered states. It's simply a matter of probability. When a system of particles changes, it's most likely to move from one disordered state to another disordered state (large objects, like cars and bicycles, are exceptions because they operate as single entities). Time is the consequence of the powerful tendency toward disorder, which physics calls *entropy*.

Several of my monks recently constructed a sand mandala of the Buddha of Compassion at a local temple. I spent hours watching them. I was enthralled at their patience and precision. Five different colors of sand grains were poured from narrow brass funnels, often a few grains at a time, to build up a complex geometric pattern. After three days they were finished. Chants and prayers followed. Then they swirled the mandala into a muddy brown pile. It tooks just seconds. No amount of swirling could ever reverse time and reproduce the mandala. In the American Southwest where I live, the Navajo and other tribes use sand paintings for healing purposes, after which they, too, are ritualistically destroyed.

In class now, I use a far more rustic example of the arrow of time and disorder. I announce, "A monk has farted." After Tenzin translates, eyebrows around the room rise. There's genuine shock, not at the monk, but at my raising the subject, or possibly at my intention of identifying the guilty party. But I assure everyone that it's alright. My point is that if we wanted to know the culprit we have a means. Seconds after the event, the hydrogen sulfide molecules are concentrated next to that person. They're ordered in one place. But we must act quickly, because with time those noxious molecules will mix in the air of the room. If the monk in question is lucky, time will save him from the ignominy of being fingered. His sulfide emission is governed by disorder, or entropy.

Actually, *disorder* is too colloquial a term to use for entropy. According to physics, a more accurate description is this: entropy is a measure of the number of bits of information needed to describe a system. More bits mean more entropy. I remind the monks of our discussion the previous week about information. To say whether there's a molecule in the room takes one bit of information: "Yes, it is," or "No, it's not." To specify the position of any single molecule we imagined dividing the room in half and half and half again, successively defining smaller and smaller regions of the space. We estimated it would take 120 bits of information to specify the exact position of any molecule.

Once our monk's sulfide molecules begin to spread around the room, it will take more bits of information to describe their positions. Entropy has increased. Put another way, the sulfide molecules can disperse across a room in a large number of equally probable ways, but they have only one way to be in a single place. Sad but true, we can trace the arrow of time back to our guilty monk.

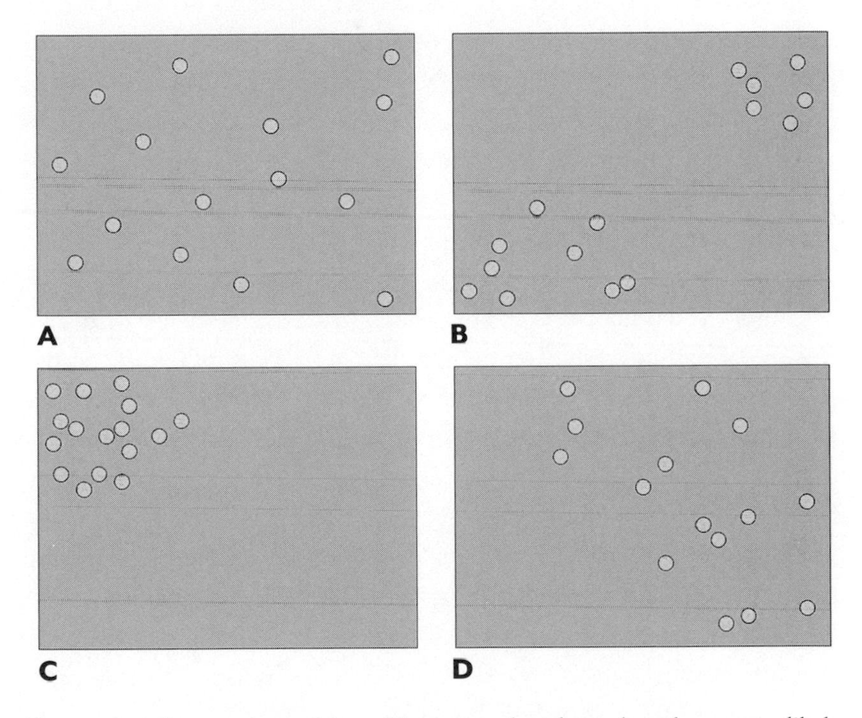

Figure 3.1. Microscopic particles subject to random forces have have many likely (A) and few unlikely (B, C, D) arrangements. Disordered arrangements dominate after many interactions, giving time an arrow.

After the morning break our discussion of time becomes general. It's freewheeling and fast-moving. Another of my regular translators is Karma Thupten (who is no relation to Thupten Tsering). He is clear and animated, but he can barely keep up. Every example leads to penetrating questions. Usually I'm struggling to provide satisfactory answers. Karma helps me clarify for the monks that the arrow of time doesn't work everywhere. It's absent in the micro world. Subatomic particles interact as if time has no meaning. Their reactions go forward or backward with equal facility. If that's the case, then why do large numbers of particles have entropy? What's the "magic" number of particles above which there's a statistical trend toward disorder?

In some of my answers I'm forced once again to use scientific terminology, or the jargon of experts. For example, I talk about the coupling of gravity and time in general relativity. Time slows down in strong gravity and freezes at the event horizon of a black hole. Ah yes, the enigmatic black hole. All around the room, hands go up and eyebrows furrow. But there's no time to explain right now. My rhetorical questions are difficult for Karma to translate, but I keep going in response to their fascination. Whose clock is to be believed if time depends on gravity? Does time get tired when gravity is strong? What happens to time inside a black hole?

A more tractable topic is how the arrow of time helps us decide the age of the universe. Astronomers put it at 13.8 billion years. But how do they know this? For example, do they presume a year for the whole universe is equal to a year for the Earth? Is time a duration or a separation between events? The questions pile up; I gamely keep lecturing. Physicists smugly define a second as the time elapsing during 9,192,631,770 cycles of radiation produced by an energy transition between two energy levels of a cesium-133 atom. What if you don't have a cesium atom handy? And I don't mean in your toolbox at home. In the early universe, how would you have measured time before there *were* atoms? Is time somehow human? Does it take a sentient brain to create the singular sense of surfing a "now"?

We're out on a limb, so I bring us back. We stick with the cesium atom but must ask: What clock does its oscillation follow? Could we subdivide time into smaller segments than the ultra-quick pulse of a cesium atom? If the division becomes infinite, then time must not have any duration. It reminds me of the Greek philosopher Zeno, who showed us the famous paradox of a turtle racing a rabbit, where the turtle wins after dividing the rabbit's space infinitely. By now my head is swimming. In a trackless and timeless fog, I wave my arms in submission, and we break for lunch.

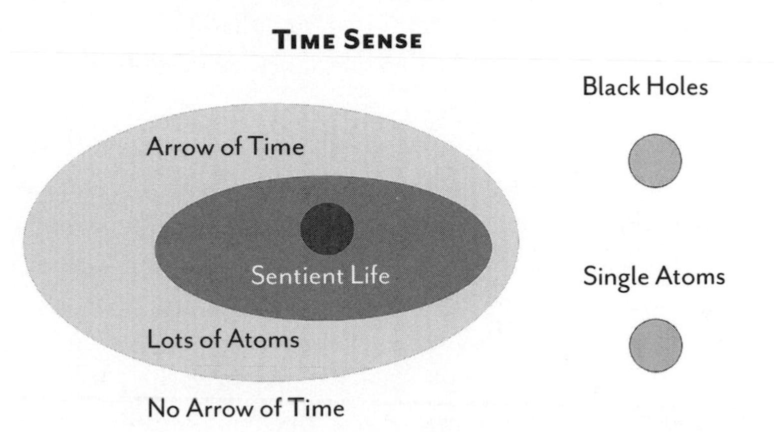

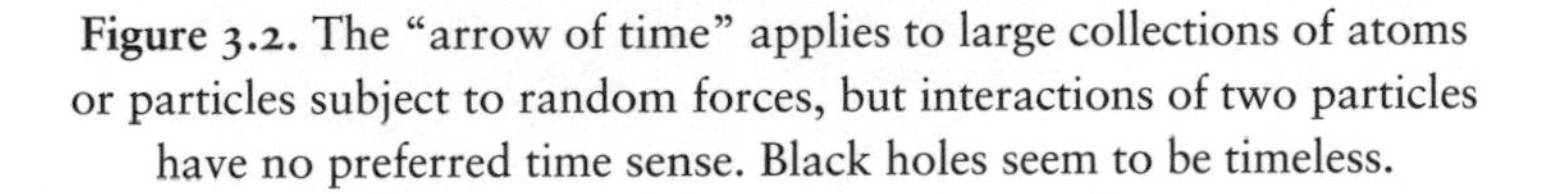

Figure 3.2. The "arrow of time" applies to large collections of atoms or particles subject to random forces, but interactions of two particles have no preferred time sense. Black holes seem to be timeless.

Speaking of time, I've run way over my allotment for this session. I've made us all late for the meal hour, and as I arrive in the staff dining areas, the translators and other teachers are just leaving. Bryce and Lhakdor are still there, talking about logistical issues with the program. Bryce asks me how it's going. Fine, I say, the monks are great. But I'm having some trouble rounding up equipment for demos. The photocopier is still broken. Some of my evening sessions had to be dropped, so the monks worked on their journals instead. I'm worried I won't be able to cover all my material. Lhakdor looks at me balefully. He apes a surfer drawl, "You need to chill, man." Then in his normal voice: "It is a terrible thing to have a monkey mind." With that, he rises, tosses the loose end of his maroon robe over his shoulder with a flourish, and exits.

I've been nailed. Buddha described the human mind as being crowded with drunken monkeys, clamoring, screeching, and carrying on endlessly. He showed his students how to use meditation to still their monkey minds. At this point in the program, I'm clearly the monkey.

Meanwhile, Bryce slouches in his chair and nurses a coffee as if he has all the time in the world. He's the glue and the animating force of Science for Monks, but I still don't know that much about him. Mercifully, Bryce doesn't dwell on my monkey-mind problem and instead is willing to tell me his story.

Bryce is the son of two scientists, but he also has the easy vibe of someone born and raised in the Bay Area. While doing an environmental science master's degree, his roommate left some books by the Dalai Lama on the coffee table, the result of taking a class taught by the noted Buddhist scholar Alan Wallace. Intrigued, Bryce took a class from Wallace on science and religion. He laughs as he recalls being the only nerd in a room of religious studies majors. Wallace became a mentor, and with Wallace as a reference, Bryce took his first trip to India to learn about the Tibetan monks in exile. It was 1999, and he was in the right place at the right time. The Dalai Lama had just conceived of the Science for Monks program.

His Holiness had directed Lhakdor's predecessor in the Tibetan Library, Achok Rinpoche, to put together a long-term project to teach monks science. Alan Wallace suggested Bryce as a smart, energetic person who could contribute to that goal. Unfortunately, Bryce had unknowingly missed that particular introduction. Even though he was in India, his visa had expired, so he had to return to the United States. A few months later, Wallace was translating for the Dalai Lama at a Mind and Life meeting. Wallace saw Rinpoche and asked him about Bryce: "What did you think of that young guy I sent you?" Rinpoche looked puzzled and said he didn't know Bryce. He'd clearly dropped the ball. The Dalai Lama leaned into the conversation. "You should find that guy!" he said.

Bryce was now on everyone's radar. When he returned to India, people in Dharamsala kept approaching him. "You need to go and see Achok Rinpoche," they said. He figured something was up. The library director knew little about science. In turn, Bryce was young, not a teacher, and knew little about Buddhism. But Bryce had an abundance of enthusiasm so he was put in charge of the new program. With a mixture of judgment, intuition, and trial and error, he developed the philosophy and the structure of Science for Monks.

At my prompting, Bryce explains that he's not a Buddhist despite all the time he's spent studying it and working in India with Tibetans. He says, "I admire the ideas and the philosophy, and I'm sure it has perceptive things to say about how to live your life and cope with the human condition. But I'm not on board with reincarnation and a few other aspects." I nod and say, "That's about where I am, too."

We're probably both sympathetic to Buddhism because, when it comes to the dialogue between religion and science, there aren't many attractive options where we've lived and studied in the United States. For example, Bryce did advanced studies in Texas and I work in Arizona, states where conservative Christians are active in protesting some key ideas in science, especially the Darwinian theory of evolution and the ancient age of the universe.

Buddhism has supernatural precepts, but it has no problem with evolution and ancient time. Of course, many groups in a diverse Christian "universe" do accommodate science quite well. In Arizona I know the cadre of Jesuit priests who are PhD astronomers with the Vatican Observatory. One of them is a relativity expert who works on the big bang. I tease him that his "God of the gaps" is squashed into the first fraction of a second since science owns the rest. To be fair, he could tease back that my "quantum magic of the gaps" fills the same space.

In the end, I think tabloid reports of the so-called science-religion wars are overstated. Unfortunately they can be deeply political, hinging on questions of public money and public education. People of faith who are scientists are no less logical and rigorous in their professions than the famous scientists who proudly declare themselves atheists and even oppose all religion. Here in northern India, the science-religion dialogue is invigorating and respectful.

Bryce and I finish our coffee. Our paths first crossed with his phone call out of the blue in 2007. Sitting across from him now, with the smell of cumin and coriander wafting in from the kitchen and the sound of chanting monks and barking dogs floating through the open window, I imagine it couldn't have happened any other way.

The Expanding Universe

Someone is on a balloon rampage, inflating with intent to burst. In our claustrophobic classroom in India, the *bang!* of popping balloons makes the hair stand on the back of my neck. In fact, it's four bangs in a row. Many times in my university classes back home we've inflated balloons to demonstrate what astronomers call the Hubble expansion. Occasionally, a student overfills a balloon and it bursts. But this seems more than innocent miscalculation, and on investigation, my suspicions are confirmed. One of the ringleaders is a monk named Jigme Gyatso. He's the monk who thought of a cloud but failed to beat the smarmy computer in our game of twenty questions. Jigme is quick to laugh and always in playful spirits. He seems to think it's fitting that a universe that began with a bang should end with one, even four in a row.

I can't be hard on Jigme, especially since he recently shared his story with me. In Tibet his family was nomadic. They lived in a Mongolian tent, or yurt, a portable round construction of bent sticks insulated with sheep felt and yak hair. He tended the flock in his bare feet. In winter the tent was surrounded by a low wall of yak dung for extra insulation. Yaks are über-animals for native Tibetans. They provide milk and hair while alive, and when they die, everything gets used—hair, meat, fat, organs, bones, and hooves. He remembers his family mourning for days when one of their yaks died.

Jigme told me about their New Year celebration, when tall dung piles were made for smoke offerings in front of the tent. Other kids teased him because his family only made one dung pile, even though his father reassured him it was special. "Before I came to India, I thought that objects such as bicycles, fans, and sewing machines made up science," Jigme told me. "After coming to India, I considered that cinemas, video cameras, trains, and cars made up science, so only the objects changed. Now I realize there are different disciplines in science, and science has made tremendous contributions toward solving problems in the world.

But whatever knowledge it might be, it can be used in a productive or a destructive way."

Despite a bit of Jigme balloon popping, we still have plenty left, and they're going to teach us about the Hubble expansion, a description of how the universe grows in size, as discovered by the astronomer Edwin Hubble. In the 1920s, Hubble made twin discoveries that transformed our view of the universe. His first involved the nature of "nebulae," which are small, fuzzy patches of light in the night sky. Astronomers who preceded Hubble catalogued hundreds, even thousands, of fuzzy objects in the night sky. At that time a debate raged. Were they merely gas and dust clouds in our own Milky Way? Or were they remote stellar systems far beyond the Milky Way? In the nineteenth century, the preferred theory was that the Milky Way galaxy *was* the universe.

Now enters Hubble. He has the best telescope on Earth, equipped with a hundred-inch mirror and stationed high on Mount Wilson in southern California. With this, he's able to decide the issue: one particular nebula, Andromeda or M31, is far outside the Milky Way. Hubble leverages earlier work by Ernst Opik and Henrietta Leavitt,

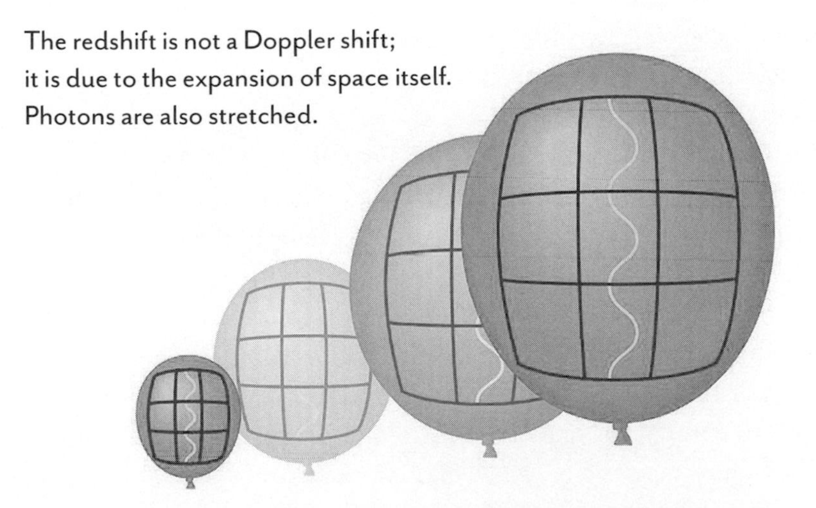

Figure 3.3. The surface of an inflating balloon acts as an analogy for the expansion of space. Beads glued to the balloon would separate like the galaxies do and radiation wavelengths increase as the balloon expands.

but unfortunately he doesn't give them proper credit. After another decade of observing, Hubble is able to expand the size of the known universe by a factor of a thousand. And of course, as the monks know, there seems to be no end to the real size of space.

Hubble's second contribution was equally momentous. He built on the work of another astronomer, Vesto Slipher, who noticed a curious phenomenon: most galaxies had "redshifts"—spectral shifts toward redder wavelengths—which meant they were receding from us. Hubble refined the measurements and added dozens more. He also saw that recession velocity correlated with distance. The more distant a galaxy, the faster it's moving away. If everything is moving away from us, do we stand at a central point? Does this mean Copernicus was wrong? Are we at the center of the universe?

The answer will be supplied by our large party balloons. It's quite a sight as the demonstration begins: three dozen monks in maroon robes inflating pink, blue, yellow, and green balloons. They partially inflate the balloons, then draw ten galaxies scattered across their surfaces. They label one as the Milky Way and use a tape measure to measure the distance to the other nine. Then they inflate the balloons twice as large, and measure the same nine distances. The resulting numbers convey the essence of Hubble's result. Other galaxies are all moving away from the Milky Way, and the most distant galaxies are moving away the fastest. The monks quickly see the point: it doesn't matter which galaxy is labeled as the Milky Way. When we look at any galaxy on the balloon surface, each has the point of view of everything receding from it, as if that galaxy was the center of the universe. This toy universe has no discernible center.

Robert Frost once said, "All metaphors are imperfect, and that is the beauty of them." Like most analogies or metaphors for complex physical ideas, the balloon has its strengths and weaknesses. As a strength, if you accept that the universe is limited to the surface of the balloon, it shows how every point can be moving away from every other point. It also shows how space could be finite yet unbounded since the surface area is easy to calculate but there's no edge. The "fabric" of space is stretching just as space-time is in our universe. The balloon also shows how a wavelength can grow. When I draw a wiggly line on a balloon, it grows with the balloon surface. This illustrates a wave shifting to redder, or longer, wavelengths.

But the balloons reach an illustrative limit. This rubber object is expanding into a surrounding room. In the real universe, the three dimensions of space need not be expanding into anything. I can see the monks are taking this on faith (and maybe I am, too). Also, the galaxies drawn with a pen on the balloon expand with it, which is not what happens to real galaxies. Real galaxies are held together by the gravity of the stuff they're made of and so are exempt from the cosmic expansion. It would have been more realistic if we'd glued beads onto the balloon surface.

Since an expanding universe is a little unsettling, I reassure the monks with a list of things that are not expanding, including the Milky Way, the other 100 billion galaxies, the constellations, the Sun, all other stars, asteroids, the Earth, all other planets, our bodies, and, regrettably, our minds.

We've gathered two sets of measurements from the balloons, one when they were the size of a grapefruit and one when they were the size of a soccer ball. The monks work in groups and make averages of their distances to nine galaxies at two times. Then we combine the results from their parallel balloon universes, and a single graph emerges on the classroom whiteboard. There's a nice linear correlation. We're seeing what Hubble saw back in 1929. The slope of the best fit to the data is called the Hubble constant, the current expansion rate of the universe. For a smooth and linear expansion, this graph can be used to infer the size of the universe, or the distance between any two galaxies, now and at any time in the past and future. That's pretty cool.

As far as I'm concerned the experiment is complete, but around the room I see balloons continuing to grow until they're ominously large, the material stretched to near transparency. *Bang!* The small room and bare walls and floor amplify the sound. *Pow!* Even though I'm expecting another one, I jump. *Boom!* Jigme may have been the original instigator, but the contagion has spread. The demise of each universe is accompanied by peals of laughter. I gesture my submission and the end of the experiment, and the monks loft the balloons into the air and bat them around. Our classroom turns into a riot of color.

There's life after the balloons. They let us calculate the rate of expansion of the universe. With this rate in hand, we can turn the clock back and speculate what the universe was like long ago. The universe has been expanding for billions of years, so in the past the galaxies were all closer together. Similarly, the waves of radiation were scrunched to shorter wavelengths, so they had higher energy. In summary, the universe used to be smaller, denser, and hotter.

Our modern model of the expanding universe lays it out in three phases, each one corresponding to the dominance of one ingredient. Until about ten thousand years after the expansion started (at the big bang), radiation was so intense that nothing could congeal in that white heat. Then gravity began to exert its grip, and until about 5 billion years ago, the expansion rate was steadily slowing. In the third, most recent phase, a mysterious entity called dark energy became dominant. It was always present in the vacuum of space, but it asserted itself as matter thinned out over time. Dark energy pushes on the fabric of space-time, acting as an accelerator of the expansion, even as dark matter, by the action of its gravity, tries to act as a brake. In this most recent phase, the white-hot radiation from the beginning has thinned out and dimmed. The space between galaxies is absolutely frigid.

Tracing back the expansion yields important insights. I remind the monks that distant light is old light. We see distant galaxies not only as they were when the universe was younger, but their radiation was emitted when the universe was smaller and hotter. Looking back half the time to the big bang we see galaxies whose light was emitted when the universe was two-thirds of its current size. And looking back 80 percent of the time to the big bang we see galaxies whose light was emitted when the universe was one-third its current size. At the limit of our vision, across 95 percent of the time to the big bang, galaxies' earliest light comes to us from a time when the universe was seven to eight times smaller and hotter than it is today. The light of some galaxies was emitted before the Earth even existed, I tell the monks. We humans arrived late in the game, but luckily we invented telescopes to reach back and "see" the ancient light.

If Hubble hinted at expanding space, Albert Einstein gave us a theory of the fabric of space-time: the general theory of relativity, which accounts for all changing motion, caused by gravity or any other force. At first, Einstein didn't accept Hubble's observation of universal recession of galaxies that space expanded, but he finally gave in. When Einstein looked at his equations, he realized that they had expansion "dialed into them." Thereafter, general relativity has been used to calculate the expansion rate at any epoch. Eight billion years ago—a look-back time that's easy to see with large telescopes—any two points in the universe were moving apart at the speed of light. The light from galaxies more than 8 billion light-years away was emitted when they were moving away from us faster than the speed of light. Yet we can see them.

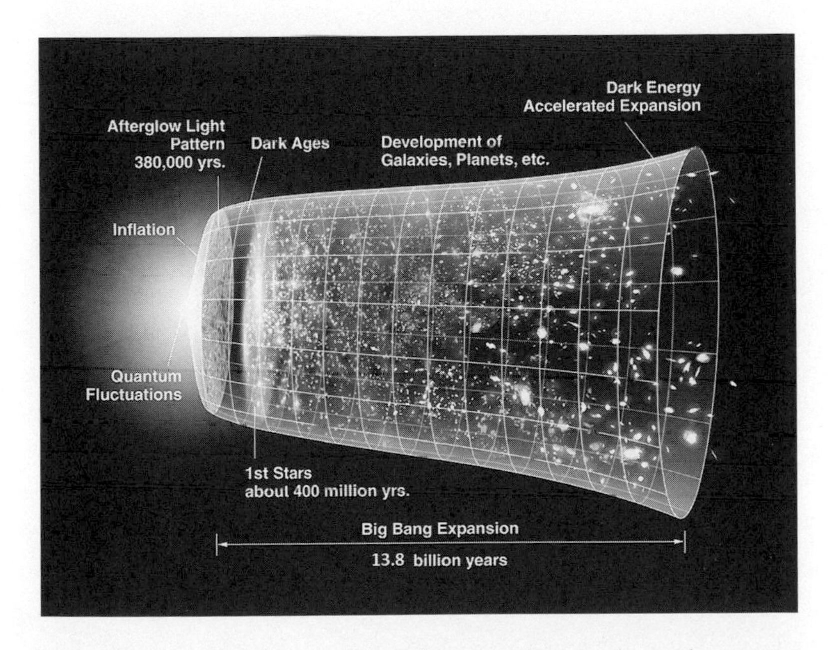

Figure 3.4. The universe had a very early phase of extremely rapid expansion, or inflation, followed by deceleration caused by dark matter and then acceleration caused by dark energy, both poorly understood.

I ask the monks, "How can that be? Didn't Einstein say that the speed of light is an absolute limit?" I pause to let this sink in. I have the monks' total attention. They're keen to know the answer to this conundrum.

The answer takes us back to Einstein and his earlier version of relativity. His "special" theory of relativity hinges on a constant speed of light in the universe. Special relativity deals with all the protean situations on terrestrial or solar-system scales where light signals are passed back and forth. For example, it explains how clocks in different regions of space and moving at different speeds can nevertheless be synchronized so we understand how they're telling time. I ask the monks, what's the difference between "general" and "special" relativity in regard to an expanding universe? With some nudging, we get there. Both theories say light travels at three hundred thousand kilometers per second and nothing can go faster. But in general relativity, there's no speed limit for the universe itself. It can expand as fast as it wants. Light does its best to keep up!

In other words, for 5 billion years the expansion rate of the physical universe was faster than light. Gravity slowed matter down. Thus, the physical universe must be larger than we can see, since "seeing" requires light. From some regions of space, light has never been able to reach us. Meanwhile, light crossed slower regions of expansion, and they are what we see. As time passes, we'll be able to see more, and that's good news for astronomers. The observable universe gets bigger every day. The bad news: every twenty-four hours it only gets bigger by 26 billion kilometers in every direction. That sounds like a lot but it won't let you see new galaxies unless you wait for millennia.

There's another bizarre consequence to the rapid early expansion. We might imagine that if the universe is 13.8 billion years old, we can see 13.8 billion light-years in any direction. In fact, our reach is larger. The distance to the edge of the observable universe in any direction is about 46 billion light-years. And the true size of the physical universe? We don't really know. It might be vastly larger, trillions upon trillions of light-years in size. So the estimate of 100 billion galaxies we reached yesterday is low. There might be many more galaxies and stars and planets (and living beings) in regions we can't see and may never see.

At a break, my translator Karma and I huddle to figure out how to physically demonstrate cosmic expansion better, since we've only spoken abstractly so far. The best we can do to show the photons in motion is to face each other while receding from each other, but also tilt toward each other, with arms windmilling as if going forward. It's definitely a vaudeville show and a pale version of Michael Jackson's famous moonwalk. But it seems to come across as a physical demo. As the expansion rate slows and light gains traction, Karma and I act that out by starting to approach each other.

Now we add complexity by considering two galaxies as well as the light that they emit. For galaxies we choose the ever-reliable Nyima and Gelek, who have proven themselves in the past as capable celestial

Galaxies are all moving away from each other, so every galaxy sees the same Hubble expansion, so there is no center.

The cosmic expansion is the unfolding of all space since the big bang, so there is no edge.

We are limited in our view by the time it takes distant light to reach us, so the universe has an edge in time not space.

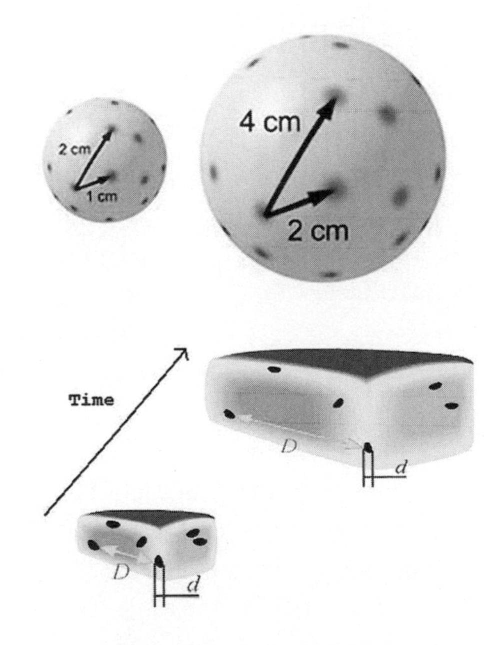

Figure 3.5. An observer in any galaxy would see universal recession, so there is no center. Analogies to help convey this are an expanding balloon (2D) and a raisin loaf baking in an oven (3D).

objects. Now they are in the early universe, moving rapidly away from each other as space expands. They'll move apart rapidly at first and then slower, to mimic the decelerating universe. I ask for two more volunteers to be the photons emitted from each galaxy. I give Dawa and Jigme boldly colored photon cutouts recycled from our light-travel-time demo. Their job is harder. As they leave the "galaxy" they must watch its changing motion to maintain a constant speed of separation. Even as they leave their galaxies at the speed of light, the photons will initially be dragged apart by the rapid expansion of space, but with deceleration they will eventually cross paths and arrive at the other galaxy. They will be "seen."

It takes several dry runs to get the astrophysics and the physicality right. At the third attempt, the demo works beautifully. The galaxies flee, dragging their photons with them. They slow down, which lets the photons approach each other and converge on the other remote galaxy, crashing into it at the speed of light, or at least the speed of monk. There's comprehension amid the hilarity around the room at the slapstick spectacle, so it's been worthwhile. Karma and I take our bows with a flourish.

Karma is a member of the crack team of translators in the Science for Monks program. My primary translator has been Tenzin Sonam, whose precise, nuanced, and skillful conveyance of my English is backed by his degree in engineering. He has been augmented by Geshe Nyima Tashi, who is a member of my class, by a new translator called Nyima Gyaltsen, and by occasional sessions with our female translator, Tenzin Paldon. Since her specialty is life science, she's working almost exclusively with Gail Burd, our biology professor from Arizona. Then there's Karma, the wild card in the deck. Karma assists me about a third of the time. He is part of the Tibetan enclave, but he's not a monk. In time, the plot will thicken: he will marry Paldon, forming the first husband-and-wife team under the canopy of Science for Monks.

Karma is a sly joker. Short and slightly built, with striking brown eyes and a smirking smile, his energy is restless. I've noticed his rapport with the monks. He hangs out and chats with them after class. Like many Tibetans, he's unafraid of physical contact. He often wrestles with the younger monks or drapes his arms over their shoulders or walks with them arm in arm.

I enjoy Karma because he keeps me slightly off-balance. We trigger each other's sense of humor. Once, I was explaining the concept of curved space-time. I paused for his translation and there was a short burst of Tibetan. I gave him a stare that said, "That's all?" The next day, when I made a simple announcement, he translated for at least a minute and elicited laughter from the monks with what should have been mundane information. I shot a glance at him, thinking, *What on Earth were you just saying?* Then I saw the twinkle in his eye; he was messing with me. But he exercises his guile sparingly; most of the time he creates a remarkable verbal bridge between English and Tibetan.

Big Bang

This afternoon the air is baked dry by the Sun blazing overhead. In the classroom after lunch I pull the red drapes to keep out direct light. But the air is oppressive; even the flies are lethargic. It's the right day to talk about "a day without a yesterday." That's how the Belgian Catholic and part-time cosmologist Georges Lemaître described the big bang. On the day of the big bang, so to speak, our currently large and diffuse universe was crammed into a space no larger than the head of a pin. We've considered how many galaxies could fit on the head of a pin and the answer is all of them. This was about 13.8 billion years ago, and three kinds of observational evidence suggest that the universe began hot and exceedingly dense.

The first is Hubble's discovery that every galaxy is moving away from every other galaxy. Run this "video" backward and you project a time when they were all closer together. As to the heat of the early universe, it behaves like a gas. If you compress the air in the cylinder of a bicycle pump by pushing hard on the plunger with your finger over the valve, the air in the cylinder will get noticeably warmer. The early universe was so compressed that it was unimaginably hot.

I knew I wanted to use bicycle imagery somehow in my teaching. Since I've been in rural India, I've realized that it's the common mode of transport. For our lesson, I borrow a pump from the man who runs a small food shop built into a small garage on our campus. His name is Yoshi, and I've occasionally bought candy and peanuts from his stall. Yoshi has henna-stained hair and tobacco-stained teeth, and he speaks very little English. We smile at each other a lot, and he seems to understand that I only want to borrow his pump. The monks take turns pushing the plunger, compressing the air, and feeling the temperature of the cylinder grow hotter. That's the early universe, hot because it's compressed.

The second piece of observational evidence for the big bang is the most convincing. It's the ubiquitous bath of microwaves that fills the universe. They were discovered serendipitously in 1965, seen with equal intensity in every direction of space. More recent satellite measurements show that they're not perfectly uniform, but have tiny ripples or fluctuations in temperature at a level of 0.001 percent. I hand out images of the microwave sky made by this NASA satellite. The variations are shown as red and blue speckles—red for very slightly hotter than average and blue for very slightly cooler than average. The average temperature is 2.7 Kelvin, within spitting distance of absolute cold.

We break slightly early for tea, and I walk into the oppressive heat and down the dusty road to Yoshi's small shop. The ten-by-twelve-foot cinder-block building has a tin roof and one wall missing. Over the counter are rows of packets of potato chips with flavors that show the British influence on India: prawn cocktail, salt and vinegar, chicken curry, oxtail, and cheese and onion. Inside the display case are forlorn bars of nougat with flies hovering above. Yoshi lies on a rusted cot on his side watching a small black-and-white TV set on a nearby chair. Grainy, flickering images show a locally made soap opera. Just what I need. He's reluctant, and he gives me the Indian head waggle several times. But I appease him with a winning smile and a major purchase of Mars bars.

Back in the classroom, after I've plied my students with snacks, I can see that I still have to persuade them that the big bang microwaves are everywhere, even in this room. The radiation was once hot, gamma-ray photons, but now it's been stretched by cosmic expansion into gentle, cool microwaves with an intensity of 10^{-5} watts, about a 10-millionth of a light bulb's worth of microwave energy. These low-energy photons pervade space, and that's our big-bang evidence. And, I tell the monks, you can watch them right now. I put Yoshi's TV on a low table, switch it on, and turn the dial until I find a frequency between the few available stations. We stare at the screen and the white noise. A small percentage of the blizzard of dots comes from interactions with the ever-present big bang microwaves. Dare I say it, staring at the interference is Zen-like. The monks don't watch TV, but I assure them this time it's perfectly okay, that this primordial show is truly cosmic and spiritual.

The third piece of evidence for the big bang involves helium. Helium is the second simplest atom after hydrogen. There's a suspiciously large amount of helium in the universe, about a quarter of the mass, far more than could have been created in all the stars like the Sun. Before the microwaves were released as hot photons, our universe was hot enough to fuse hydrogen into helium, something that happens in the core of the Sun. This fusion occurred for a few minutes after the big bang and then stopped when the universe cooled below 10 million degrees Centigrade. This explains why there's so much helium in the universe: it was mass-produced at the very start. For a moment, I wonder if Yoshi has a cylinder of compressed helium, since he does sell party balloons. How wonderful if this tiny merchant with red hair and faded blue pajamas could help me with all aspects of the big bang. But no, another visit to his store to assuage Paul's sweet tooth confirms that he can't help me with helium.

All this talk of hot and dense states is skirting around the issue of the very beginning. Running the clock back to time zero implies infinite temperature and density. Mathematicians are fine with infinities but physicists abhor them, and normal people aren't too happy with them either. It seems nonsensical to imagine the universe shrunk to a point. But that's what the big bang implies.

Imagine the Singularity! Now the monks, along with me, take an audible gulp, followed by a few moments of contemplative silence. Then, like a big bang, the questions come in a flurry. "What caused the big bang?" "What came before that?" "Was it an explosion?" "How

MICROWAVE BACKGROUND

The universe is expanding and it was also much hotter and denser in the distant past.

The microwave background and the helium abundance cannot be explained in any other way.

Every cubic centimeter of space contains millions of photons from the big bang. It surrounds us.

The big bang theory is supported by a web of evidence, but theory does not explain the cause.

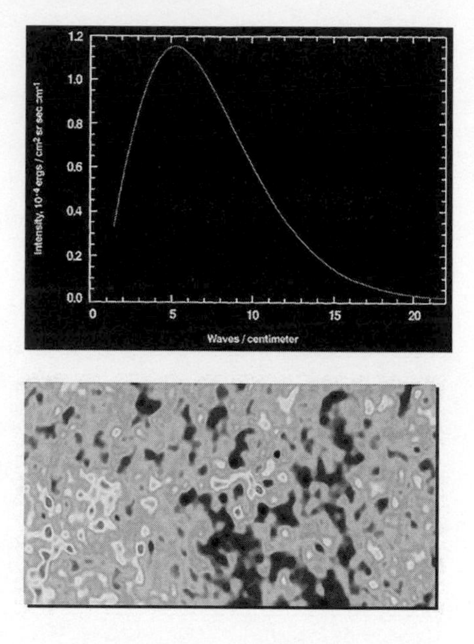

Figure 3.6. The almost uniform microwave radiation seen in every direction has exactly the right temperature to be a relic of the early hot phase of the universe (top); there are very slight variations in the temperature (bottom).

can the universe have been a singularity?" "How do we know that was the beginning?" I take shelter behind my hands in mock fright. They're great questions but I don't have good answers. As always, Thupten is in the vanguard of the questioning. With his sharp mind and connection with the cohort, I use him as an interlocutor to gauge where the monks are in their understanding.

The hardest thing for the monks to fathom is the fact that the universe can expand without expanding *into* anything. Despite the moniker, the big bang wasn't really an explosion. In general relativity, space and time are joined at the hip and must be considered together. The habit of talking about time and space as if they were as different as chalk and cheese is only possible because our universe today is old, cold, and flat (meaning we can measure it using Euclidean geometry, just as we could measure a building with three orthogonal dimensions). However, the early universe was a time when space was highly "curved." The big bang was an event where space unfolded like a lotus flower. Space *is* the flower—there's nothing outside.

Eyebrows wrinkle around the room at the idea of a spontaneous creation of a vast, evolving universe from a tiny kernel of space-time. Monk credulity is being strained. Sir Fred Hoyle, the English astrophysicist who invented a rival theory, gave the big bang its name and he intended it disparagingly. Hoyle believed that the universe didn't need a beginning. He suggested we see matter slowly and steadily being created in the voids between galaxies and this is what makes space expand. In his model, called the "steady state," the universe is eternal and it has always looked more or less the same. Hoyle thought that instantly creating the stuff of billions of galaxies from nothing was as ridiculous as any of the world's creation myths.

I can't deny the sheer audacity of the big bang theory. But it's our best explanation, plausible because it has passed so many tests and explains so much evidence. I tell the monks that cosmologists are reaching for additional theories to describe physical conditions at the instant of creation. If they're successful, these theories might account for the big bang as a quantum event in a situation where many other universes might exist.

The wide eyes around the room tell me that they're taking this so-called quantum event to be just as amazing, even fanciful, as any ancient legend. They all remember as children how their parents told them Tibetan folktales, often with a moral to the story. When I speak of quantum events, and many universes bubbling up, like soap suds in a sink, they have every right to think I'm telling them another folktale. I've stretched their credulity to the limit.

Time without End

It's nearly halfway through the workshop. I'm enjoying myself immensely. Yet a sadness is creeping up on me that it will end. *How perverse*, I think, catching myself. Is this pessimism the monkey mind alluded to by the Buddha? Why can't I just take my first happy thought and accept it, pure and simple?

At breakfast I meet my fellow teachers Gail and Mark, and we catch up on events. The schedule is demanding so we've rarely had time to chat. Class preparation and teaching consume about ten hours a day. When that's done, I spend what's left with Paul. Bryce joins us, and he's the bearer of dark news. Another monk in Lhasa has set himself on fire to protest the Chinese oppression in Tibet. I try to imagine, but simply can't, the act of will and despair that would lead someone to take their own life in such an agonizing way.

The next class, the monks arrive in a glum mood. They huddle in groups, whispering. I'm about to ask Karma what they're talking about, but one look at their stricken faces tells me. I tell them we'll start the class at nine instead of eight today. I take the extra time to sit outside on the wet grass and close my eyes and savor the unfamiliar fragrances that lace the air.

I'm momentarily caught between two opposing feelings. On one side is sympathy for the self-immolating monk, from the perspective of a person like myself with a family and a set of intimate, human, subjective experiences. On the other side, however, is the lesson I'm about to teach in class. It's a lesson of the sheer puniness of human beings in the universe, whose size we're about the measure in one of our scale models. Does a single life matter? What is the meaning of a single immolated human in a vast and impersonal cosmos? As human beings we have empathy, so it has to matter a great deal.

At times like this I'm glad I only teach astronomy and not what the philosophers and theologians call "theodicy," our way of coping with death, loss, and evil in a universe that seemingly doesn't care a drop (or, for believers, a universe with an omnipotent God who allows the suffering). What of our lifetimes: that arc of hopes and dreams and opportunities taken and not taken? It's a minuscule fraction of the time span allotted to stars and galaxies and the universe itself. How tiny? Let's find out.

The monks push their mats sideways so that a long wall of the rectangular classroom is free of obstructions. I've measured it previously to be forty-five feet, or about fourteen meters, which makes for a convenient timeline where one meter is a billion years. There are no too-cool-for-school monks here, and all of them take up volunteer tasks. I hand out items that identify various landmarks in the history of the universe, and ask the monks to position themselves suitably along the wall. Our chief landmark will be the big bang, and for this I tap Dawa. He's suitably compact and full of energy. Dawa gets a piece of fluorescent yellow paper cut into a jagged explosion shape, and he wears a pair of Ray-Bans, for obvious reasons. I place him in the front corner. All the way across the room, at the other end of the wall, is "now," I tell the monks.

Then I hand out the rest of the items: a picture of a spiral galaxy to represent the formation of the Milky Way, a model Earth to represent its birth, a rock with algae on it for the first life on Earth, a fern frond collected on a walk the day before for the first plants, a dead scarab beetle I found in my room the day I arrived, a plastic snake and a plastic mouse to stand in for the earliest reptiles and mammals, a magnolia blossom from just outside the monastery for the first flowers, a model action figure to play the role of the first humans, and a small "happy Buddha" statue I picked up at the airport while waiting for my flight to Dharamsala.

I let the monks loose and then melt back to the opposite wall to watch. Dawa has the easiest assignment and he leans serenely in the corner wearing his shades, with the big bang icon tucked in the waistband of his robe. He seems quite comfortable being the creation event for everything that follows. The monks holding the other items try to find their correct places along the wall, showing the order of events in cosmic history. There's shuffling and to-and-froing. Laughter rises in the peanut gallery. Spectator monks offer unsolicited advice, and Geshe Nyima comes out and exercises his seniority by yanking another monk into the correct position in the lineup. The noise subsides.

They've positioned themselves more or less evenly along the wall, with slightly more concentration of events near the "now" corner. Several of the monks are in the wrong order. I move forward and give the correct chronology, gently shifting monks as I speak. The Milky Way forms not long after the big bang, so it's a yard away from Dawa. The Earth forms two-thirds of the way along the wall, and the rock with the algae is only 2 billion years later, or six feet further along the wall. I gesture that the monks holding plants and animals should move much closer to the corner farthest from the big bang. I herd all five of them into the corner and start leaning on them with my forearms, playfully trying to crush them into the two feet nearest the corner, where they belong in this scale model. Other monks lean in too. There's shouting and some raucous laughter, and everyone piles into the rugby scrum of late evolutionary life.

A billion years is a meter, I explain. So all the development of life on land, the proliferation of plants and animals, takes place in a couple of feet. And what about you? I gesture to Gelek and Nyima, who hold the early human and the Buddha. They look at each other quizzically but realize that the human part of the cosmic saga is very recent. In fact, as becomes clear when we work it out on the board, humans are located two millimeters from the far wall of the room. The life of the Buddha would be a microscopic two thousandths of a millimeter from the wall—less than the thickness of a human hair, a sliver of enlightenment in the cosmic pageant.

Imagine a cosmic calendar, where the universe began on the first of January and now is the stroke of midnight on the thirty-first of December. This compresses 13.8 billion years of cosmic history into a terrestrial year, so analogous to scale models of space it's a 14-billion-to-1 scale model of time. In this shrunken version of history, the first stars form in mid-January and the first galaxies form in early February. The universe has plenty to do—expanding a few hundredfold and forming billions of individual galaxies and tens of thousands of galaxy clusters—but the next event of interest to us is the formation of the solar system and the Earth. This comes in early September.

Life forms fairly quickly in this cosmic year, by late September, but it isn't until near Thanksgiving that life moves toward complexity by the innovation of multicelled organisms. Sex was invented on the eighth of December; even though monks are celibate, it's imperative to put in on the calendar. This shuffling of genetic material (that is, microbial sex) accelerates, as does the growing sophistication of biological functions. In mid-December there was still no life on Earth larger than the head of a pin, so all the diversity and proliferation and plants and animals occurred since then. Apes don't appear until late morning on the thirty-first of December. All of human civilization happens in the quarter hour before midnight.

Glancing at a real watch on my wrist, I realize we're running late and have spilled into the class time for Gail and biology. When we gather again, I'll introduce the monks to the ultimate events on the cosmic calendar. I've saved the worst for last.

For this sense of pending doom, I quote the eloquent Irish poet William Butler Yeats: "Things fall apart, the center cannot hold." Thanks to the arrow of time, we all face death as an irreversible outcome. Only the manner and timing are up for grabs. Maybe we should take solace in the grand demise of the universe and be amazed that we can comprehend something that so dwarfs our time in the limelight. Throughout human history we've also taken solace in a number of metaphysical systems that give us hope for a future life, or at least say that our lives amounted to something; that all the effort is not for nothing.

Personally, I'm agnostic on such solace-giving systems. I have an uneasy relationship with the nothingness that awaits me when the ferment of my consciousness subsides. But I try to understand how believers adopt hopeful outlooks based on a theological system. One of these systems is reincarnation, which is at the heart of Buddhist belief. Reincarnation is understood in very diverse ways even by Buddhists. I arrived here with a sixth-grade idea of reincarnation. Westerners hear the parables about people coming back as plants and animals, or if good, rising to be angelic bodhisattvas, spirits who return to Earth to help fellow human beings. Buddhism the world over does seek "good fortune" from such supernatural sources, evoking the afterlife at shrines, temples, and graveyards.

However, since coming to India I've begun to understand the more sophisticated version of the belief. Reincarnation is not about fate or the transmigration of the soul. Bad karma doesn't make you an insect in a next life, though the idea haunts me when they crawl around my room at night. Karma means the actions we take. Actions can change the present and the future. We're subject to the effects of our actions. In this sense karma is like a law of conservation of moral energy.

Another level beyond the popular notions invokes metaphysics or mysticism. In Buddhist tradition the higher and lower realms of existence involve cycles with very long duration. In the realm of "forms," or physical existence, lifetimes can stretch to over 25 million years. In the realm beyond forms, existence continues up to eighty thousand great kalpas. A great kalpa is an eon in Sanskrit, defined as a very, very long time in human terms, as we saw in the story of the granite mountain eroded by the wing of a dove. And time spans in the Buddhist version of hell? You don't want to know. The story is told of a suffering ghost who hadn't eaten for such a long time he focused on someone nearby who was about to spit, since he craved the spittle. During his wait, the closest city crumbled and was rebuilt seven times before he got the spittle.

Our departure from this physical world—our death, in other words—is quite mundane. In the light of science, death can't be avoided by chants or prayers, I tell the monks honestly. Death can be temporarily avoided only by caution and hygiene. I give them a short list, tongue in cheek, but based on science and statistics: make sure all food passes the smell test, avoid baths, cars, beds, and other places where most people die, and finally, watch your friends carefully, because people are usually murdered by someone they know.

Even following this good advice for individuals, I go on, we don't yet know a way to evade or thwart the demise of the human species. One way to attain human immortality is by a mash-up with machines, in a kind of cyborg scenario that's meat and drink to any science-fiction enthusiast and anathema to any humanist. However, the evidence isn't hopeful. All past species on Earth have gone extinct. The smart money says that humans won't be an exception to the rule. While we're the current top dogs on the planet, evolution is restless and will likely lead to some worthy successors.

Since this is an astronomy class, I tell the monks that our destruction may also come from outer space. Their eyes widen at this, and a few monks glance upward. We're not talking of an invasion from Mars. The solar system is littered with detritus left over from its formation, and occasionally we have a run-in with a space rock. The most predictable sources of mayhem are the streams of comet debris that Earth's orbit passes through each year. Most years the particles are small and the result is a meteor shower. But every thousand years or so big chunks hit the ground and cause local damage. Every few million years a rock the size of a small building hits us and decimates an area the size of a small country, and every hundred million years or so something the size of Manhattan causes a mass extinction like the one that killed dinosaurs and many other species. When the Big One comes, climate change will be the least of our worries.

Other threats come from farther afield. When a massive star dies as a supernova, it obliterates itself in a torrent of gamma rays. If one were to go off within thirty light-years, it would cause cellular damage to all land creatures and wreak havoc with the food web. An even more dramatic stellar cataclysm called a hypernova could disrupt genetic material at a distance of a thousand light-years. Luckily, violent star death is rare. Yet there's evidence that they cause miniextinctions every 10 to 15 million years.

I watch the monks for a hint of angst. They're serenely unperturbed. As students of the great wheel of existence, they know that all manner of calamity can shuffle our existences into flux. They're equally calm when we move on from mere destruction of the Earth to destruction on the level of our solar system. This demise will begin with our Sun. Although is has another 4 billion years of fuel left, the Sun will steadily burn hotter as it uses up its supply of hydrogen. This gradual warming will boil the oceans and destroy Earth's biosphere. So if we have very distant heirs, they'll either have to figure out how to live underground or go off-world.

Surely the universe itself will go on prospering? A casual observer could be forgiven for thinking that the universe is timeless. True, the galaxies pirouette and imperceptibly move farther apart, and within them stars are born and die. But in a trillion years, give or take a few billion, the cycle will be broken as more gas is trapped in stellar corpses and less is left over to form new stars. The lowest-mass stars will be the last to complete their miserly fusion phase. In each and every galaxy, the lights will gradually go out. Fade to black.

Stellar Lockdown

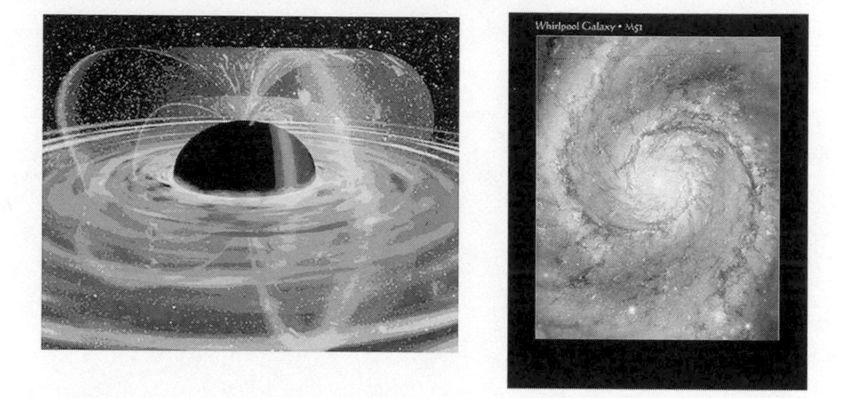

Within the Milky Way, and all other galaxies, the cycle of star birth and death will be broken in a trillion years. End states are black holes, neutron stars, and white dwarfs.

Figure 3.7. In the far future, the gas supply for new stars to form in the Milky Way and other galaxies will dry up, so a growing fraction of the mass will be in the form of dark remnants and fading embers.

Meanwhile, dark energy will drag galaxies away from each other at an ever-increasing rate. If we were around to watch, we'd see Andromeda swooning into the Milky Way's arms. This is more a gentle gravitational dance than a giant collision, because the spaces between stars are so vast. Within galaxies, physicists predict that normal matter will decay and black holes will slowly evaporate. The end result is an undifferentiated, diffuse froth of electrons, positrons, neutrinos, and photons. Physicists aren't unremitting gloom mongers; they bring a ray of hope. If life in a general sense is a process of extracting information from available energy in the cosmic environment, then biology may be parsimonious enough to persist in a decaying, accelerating universe.

The monks absorb this bleak scenario with equanimity, and it helps that I've learned a playfulness from them that flavors my story of doom with humor and imagination. Either way, their essential lightness is untouched. They also know I'm presenting speculative theory. No point in getting out of sorts due to metaphysics. They've fully accepted their role in the universe as bit players, not on center stage.

I show parallels between the metaphysics of Buddhism and Western science. Such is the case with the Buddhist "kalpa," which means eon. Buddhist kalpas come in small, medium, and large with sizes from 16 million years up to 320 billion years. Then there's a great kalpa, the lifespan of the universe: equal to four large kalpas or just over a trillion years. This is very similar to the timescale of the cosmological prediction when all stars will burn out and die. This is also when the accelerating universe will be destroyed by dark energy. During a great kalpa, the universe comes into being, is destroyed, and emptiness ensues. Then it all starts again.

No sad farewells, no tear-filled eyes. In the eons that await us, at the juncture when the universe has lost all structure, time will have no meaning, just as it had no meaning at the birth. The quantum foam that spawned our universe may have generated space-time confections and life in other universes beyond our imagination. As poet e.e. cummings said, "There's a helluva universe next door, let's go!"

During a visit to Yoshi's shop, it crosses my mind that I'd like to ride a bicycle. My little "universe" of the campus feels a bit limiting. But the condition of the roads and the insanity level of the driving cure me of that urge. Also, of the few interesting places to go, Dharamsala is at the top of the list, and that would mean pedaling uphill for ten miles. As Paul and I get our bearings, we start going to Dharamsala by taxi every couple of days. It gives us fresh air, exercise, food for thought, and a welcome diversion.

One day we're acting the tourist, checking out the gift shops and enjoying milky coffee and banana bread at a coffee shop on the edge of town. We hike a mile farther to a waterfall. According to legend, the region suffered a severe drought a thousand years ago. In desperation the king descended to the underground kingdom of the Naga, a fierce and serpent-like race. He stole water from them but they pursued him to the surface, and in his panic he spilled the water. He prayed so hard they let him escape with his life, and the waterfall is a reminder of the episode.

Back at the college that evening, Paul and I hang out in the canteen where the monks eat. Away from their home monasteries and their normal routine, they're having a great time. We talk to the few who speak good English but mostly let the Tibetan wash over us like warm waves at the beach. Being a monk is serious business, but they also like to play games of skill.

One of their favorites, played on a small square table, is Karom. The table has a rim and holes at each corner, like a cross between pool and shuffleboard. Each person takes turns flicking small disks off the sides and into one of the holes. The surface is scattered with flour so the disks will move smoothly. Several of the monks are wickedly good. Ping-Pong is also popular. The table is cracked and uneven and the net is a row of bricks, but the level of play is high. Encumbered by robes, some monks still manage to execute shots with vicious, looping topspin. At both games, they kick our butts.

My forays to Dharamsala are not always for tourism. That Sunday, I'm woken by rain pounding on the roof. When the sky opens up here by the Himalayas, it's biblical. Looking out of the window into the gray light is like looking through a waterfall from behind. The raindrops are merging on the way down into balls of water the size of marbles. They burst when they hit the dirt. Unfortunately, Paul's nausea and digestion have taken a turn for the worse. I gather his medicines and wait for the rain to relent, feeling stir-crazy. Paul finally falls asleep and will be out for a few hours, so I ask Sangey to get me a cab to Dharamsala.

I've spotted an Internet café in a dingy back alley. Inside, six computers are crammed in an impossibly small space, separated by flimsy cardboard partitions, with stools to sit on. Three of the six places are occupied by monks. I wait my turn. I pick carefully through my hundreds of e-mails. So far, my fragmentary sessions on the Internet have only reminded me of all the work I've been getting behind on. Though amicable, my colleagues are now my tormentors, peppering me with numerous requests. My day job is eight thousand miles away, but it clings to me like a viscous, sticky substance. To feel better about all this, I send a few e-mails of my own, upbeat messages to select friends. I want to give them a taste of India and my recent experiences.

As I stand to leave, I see a young monk in the cubicle next to me streaming *Family Guy*, a hip and snarky cartoon sitcom on the Fox network. Buddhists are not all as otherworldly as I imagine. I pick up some English biscuits and Cadbury's chocolate for Paul. As I might have expected, I'm on the street when the skies open. In running thirty yards to the awning of a nearby shop, I'm utterly soaked. My clothes cling to me and rivulets of water run down my back and legs. Two monks sheltering under a big rainbow umbrella point at me and giggle.

I walk toward the taxi stand, too wet to care about trying to shelter from the rain. On the spur of the moment, I duck back into the alley with the Internet café. My thoughts are fixed on a persistent theme of my time here—the contrast between a career like mine and the Buddhist belief in impermanence, as illustrated by their wistful lives and their elaborate mandalas, built up only to be swept away.

I'm here in India to teach science, which is a product of Western culture. It has advanced our health, comforts, and knowledge in ways that folk wisdom could never do. I bend under the Socratic self-examination brought on by the simple life of monks and the existential detachment of Buddhism. But I don't seem to break. How can I give up what Western culture has made of me, and abandon the fruits of the Western mind that I like so much?

I face a kind of ultimate choice, or maybe I'm being too dualistic. For now I put it aside. Still soaking wet, I log on to my e-mail, think for a minute, then set the automatic vacation reply to read, "Thanks for your message. While sympathetic to your request, I'm unable to respond and I may in fact stay here and heed the call of the maroon robe." I'm not entirely kidding.

The Forces of Nature

"MY MONKS." In my head and in conversation, I've started to use the possessive when talking about the Science for Monks students. It's also easy to talk about the monks as if they were one entity, a Borg-like collective with communal thoughts and feelings. Even Bryce says "the monks" more often than he says "a monk" or talks about any of them by name. Language is a barrier; I can only have a freewheeling conversation in English with three or four of the students.

But I've been learning about their histories and personalities, and their vocation and training in no way render them monotone. We joke about it in class. They, as monks, lead boring lives and are always meditating. I, as a scientist, am focused, humorless, and always calculating things. Such banter and joyful self-criticism is remarkably easy among my monks.

As I've learned, there's no typical monk's life, although they have the same cultural origins and have experienced similar hardships. They're all trying to follow the eightfold path, which is Buddha's practical guide to ethical and mental development; the goal is freedom from attachment and delusions. The monks tell me their stories only at my prodding. It's unmonklike to call attention to yourself. None of them presume they're special or have suffered more than others.

One of those stories comes from Gelek Gyaltsen, who we recall calculated that he eats 7 million grains of rice a year. He's very tall, with graceful hands and a huge wingspan; I watched him use his arms for emphasis very effectively during the debate the monks held after prayers for the Sagadawa festival. He heard the calling to be a monk, just as I heard mine to be a scientist. Gelek has also studied in the United States so we have common ground. Gelek begins at the beginning. He tells me he was born in Tehor in eastern Tibet:

> I remember when I was eight or nine years old my parents would send our animals off in the morning into the mountains, and it was my job was to bring the animals back in the evening. It would rain and they would hide under boulders and not come out easily. I cried a lot during those times because if I didn't bring all the animals down my family scolded me. When it got dark, me and the kids my age got really scared. There were lots of wolves, and we had to keep them away from our sheep and cattle.
>
> At the little school in our village, I don't remember studying much. Most of our lessons were in Chinese. We played in the snow in winter and hunted birds in the summer, and the teachers were just there to pass the time and they let us play or do whatever we wanted. I only went to school for two years. One day, all of a sudden, my best friend left to join the monastery. I begged my father to send me, too, but he said, "Don't be in such a hurry. You can become a monk in a day. You just shave your head and take your vows, that's it. Better you should continue your education."
>
> In our family my father had two wives. The younger one was my real mother, and the older one we called "Grandma." Grandma had a lot of influence in our family, and she was very religious. She argued with my father and won him over, so it was decided I would become a monk. I was fourteen years old.
>
> I went to the local monastery but it didn't have good monastic training, so I wanted to go to India, where monasteries are better. I secretly went to Lhasa to try and get a company and guide to take me to India, but I couldn't find any. I stayed for a year and ran out of money. I went back to

my village monastery to get help from my teacher. Two other young monks wanted to go with me, so we worked together. But the grandmother of one of them told my family and both my mothers came, crying and begging me not to go to India. I told them I'm not learning anything in the monastery here.

They took me home to discuss it with my father. He was upset. He said he never cried for me from the time I was born but he cried when I left for Lhasa, so now if I want to leave again, "Think that you don't have a father and I will think I don't have a son and that will make life easier for both of us."

He didn't speak to me for a week. I felt very sad and went back to the monastery on my small bike. One day I was trying to sleep, and there was a knock on the door. It was my mother. She said she had talked to my father, and they decided to send me to India. The next day he came, too, and they agreed to raise money for my trip. He told me not to tell anyone else in the family or they will be saddened and it will be difficult for me, too. The day before I left, they came with lots of food and drinks, and we spent the whole day talking and eating. I still feel bad that I left without telling my sisters and brother. I was eighteen years old.

When I got to Lhasa with my two friends there were many protesters and police on the street using tear gas, so we stayed inside for several days. It took us a month to get to Nepal, first in the back of a truck but mostly on foot. There were twenty-eight of us with my friend as the guide. We ran out of food several times and traded dzi for staples like rice, potatoes, lentils, and tsampa. Dzi is a special agate that has spiritual healing powers. Later we had to exchange clothes for food since the mountain people didn't have good outer wear. Sometimes we stole corn from the fields.

Crossing the mountains was hard. We passed people who had died in the snow and we stopped to say prayers for them. Luckily nobody in our group died.

After crossing into Nepal at a place called Kitali, we were caught by the police while we rested in an open pasture. They took us to the police station. They didn't give us any food but luckily they didn't take our Chinese money. They sent us on a bus to Kathmandu. We smuggled out notes saying we had been arrested on the way from Tibet and to please alert the Tibetan refugee center in Kathmandu. In Kathmandu we were in custody for a week and given no food. But other Tibetans brought us food, and a sympathetic Westerner gave us a thousand rupees. To this day I don't forget his kindness.

Finally we got to the refugee center and then to Delhi. I was sent to the Drepung monastery. I met His Holiness for the first time at the Gyumed monastery.

My older mother died before I left Tibet, and my father died five years after I left. My brother is a driver, and he's married and has three kids. One of my sisters is a nun in Tibet, and the other sister is married and is a teacher. I haven't seen anyone in my family for more than fifteen years. It was difficult, but before 2008 I could contact them. Since then, the Chinese have blocked the phones and the Internet.

I've met Westerners who think that monks are serious and always meditating. They're really surprised when I tell them I don't meditate. In our monastery we spend a lot of time studying and understanding the core Buddhist texts. We refer to commentaries by different authors and use debate to clear up misconceptions. But meditation is a private matter and we don't discuss it with others. It's also not true that we're serious and live in isolation. We have a sense of fun and we use laptops, cell phones, and the Internet. They should understand that we're all the same human beings as they are—we share the same feelings, suffering, and joy that they do.

When I think of my friends back home with limited freedom and no access to good education, I feel fortunate to have traveled so far and learned so much, and I feel sorry for them.

> If I had stayed in Tibet even in my dreams I couldn't have had such opportunities.

The Material World

Gelek Gyaltsen is using the workshop to learn about the material world. Materialism is science's greatest strength and its greatest limitation. Theories of matter have led to insight into the structure of atoms and the way they combine into the chemicals and materials that fuel the modern world. Yet those theories can't explain consciousness or the profound capabilities of the brain. They're mute to meaning. Materialism casts no light on morality or the human condition.

This is my caveat as I present the monks with the scientific ideas that seek to unify nature and explain everything from the tiniest subatomic particle to the observable universe. I pause to ask my monks: Why do cosmologists care about the tiny atom, since their interests seem to be on a larger scale? They're catching on. It's because the universe began as a kind of atom. It was once a singularity, a situation where all particles and galaxies were merged in a primordial state that was unimaginably compressed and unimaginably hot. That was 13.8 billion years ago. In the tiniest fractions of a second after the big bang, no particles even existed, and hence there was no such thing as "material." Nevertheless, materialism started in this elusive realm, and that's why physicists and astronomers are both eager to understand the subatomic world.

At the start of the universe, everything was united, not separated into particles and forces, which would come later. Culturally, humanity has intuited this unity. One of our universal symbols is the *ouroboros*: a snake that eats its tail. It's found in many civilizations across the world spanning five thousand years. It represents a primordial unity, and the Swiss psychiatrist Carl Gustav Jung identified it as an archetype of the human psyche.

In cosmology the ouroboros is an allusion to the fact that the big bang can only be understood in terms of a fundamental theory that reduces matter to its essence. And the corollary means that tests of that fundamental theory might only be possible via observations of the early

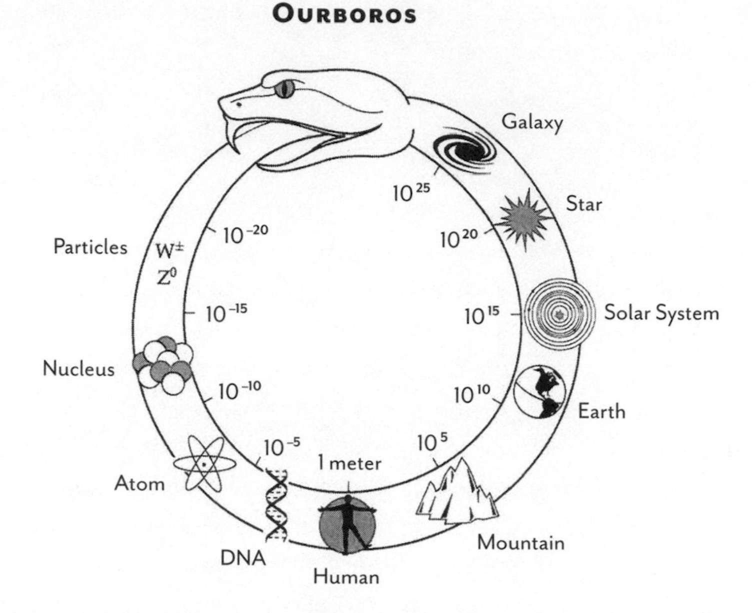

Figure 4.1. The **ouroboros** is an ancient symbol of interconnection. The snake eats its tail, and in the physical universe the microscopic and macroscopic worlds are united in the cauldron of the big bang.

universe. The Dalai Lama makes reference to this idea in the title and contents of his 2005 book, *The Universe in a Single Atom.*

How can I make atoms real for the monks? I've told them that a few billion could fit across the head of a pin, and I've shown micrographs that zoom in on the colored sand grains of a mandala until shadowy forms appear that are small clusters of silicon dioxide molecules. But that's too indirect, too esoteric.

For a hands-on exercise to probe the atom, I turn to Gelek, our tall monk with the graceful arms. I hand him a lump of flour dough about the size of a soccer ball that I've purloined from the kitchen this morning. He cradles it delicately. I say that the dough is made of atoms. He's to cut the dough in half again and again to see how far along the road to atoms he can go. We've seen from our consideration of bits of information that subdividing space can be an efficient way of homing in on the truth. Half a dozen of the monks crowd in to get a closer look. After twenty cuts, the dough is a small pale lump on the end of his finger, no bigger than a peppercorn. He shifts from using a knife to using the end of a needle. After thirty-five cuts he has to give up. The dough is so tiny it nestles in a groove of one of his fingerprints. I can barely see it. It's about 30 million times less dough than when he started—not enough to feed one of the ubiquitous mosquitoes—but it's just a few hundred times smaller, about a third of a millimeter across.

Two and a half thousand years ago, Democritus speculated about the subdivision of matter. Logically, there are only two possibilities. Either the process of subdivision can continue endlessly in an infinite process, or it reaches an end point with indivisible units of matter. Democritus thought the latter made more sense so he hypothesized atoms. It would take another sixty cuts to reach the level of atoms. Going from a huge wad that's enough to feed all the monks to a tiny nub of dough is less than a third of the way.

We get closer with some very simple equipment (atom-smashers being rather rare in rural Himachal Pradesh). The monks gather around as I fill a wide, shallow pie dish with water and sprinkle some pepper on it. The pepper's only purpose is to mark the surface. Then I let a tiny drop of dishwashing liquid fall onto the surface. The pepper flees to the edge of the pie dish as if pushed by an invisible force. The monks are impressed. But what does it mean? The dishwashing liquid has an ingredient that repels water, which causes it to spread out as it hits the surface. We work through the fairly easy math. Geshe Nyima, Jigme, and Thupten do it alongside each other on the whiteboard, to see if they get similar answers. They do.

As it hit the surface, the small sphere of dishwashing liquid morphed into a squat cylinder. Setting the volume of the sphere equal to the volume of the cylinder, the only unknown quantity is the shallow height of the cylinder of dishwashing liquid. The result is impressively small: one-hundred-thousandth of an inch, far smaller than the width of a human hair.

This is within spitting distance of atomic scales. And if we'd done the experiment in a bathtub or, better yet, let a small drop of oil spread out due to the action of light wind on a pond, we might have made a true monolayer, a layer that's just one molecule thick.

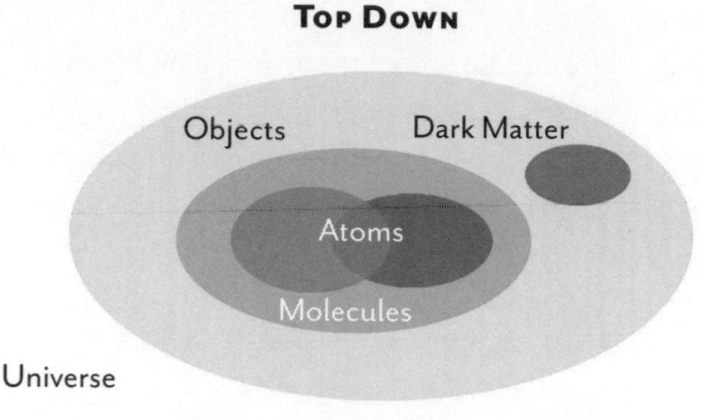

Figure 4.2. The classical or "top-down" concept of structure says that the universe is composed of objects, which are made of different combinations of molecules and atoms. Dark matter is distinct from normal matter.

However, molecules are made up of atoms, so the real "meat" of matter is on an even smaller scale. Atoms are kept apart at their normal spacing in matter by a diaphanous cloud of very light electrons, and at the atomic core—the nucleus of the atom—we find almost all the mass. The nucleus is incredibly heavy, hard, and dense. The physicist Ernest Rutherford demonstrated this with a series of elegant experiments in the early 1900s. He fired a beam of high-energy particles at a thin gold foil. He expected them to be stopped by the foil, since it was believed that atoms were impermeable. But to his surprise, most of the particles passed through the gold foil undisturbed, and to his even greater surprise, some of the particles bounced back. As he famously reported, it "was almost as incredible as if you fired a fifteen-inch shell at a piece of tissue paper and it came back and hit you."

Rutherford interpreted these results as indicating that matter is mostly empty space and that most of the mass of an atom is in a very small and very dense nucleus consisting of protons and neutrons. The nucleus is like an orange on the fifty-yard line of a football stadium, where the electrons occupy a region stretching out to the nosebleed seats. Adjacent atoms resist each other due to the electrical repulsion between their electron "clouds."

When we rap our knuckles on a table we feel and hear a pleasing solidity, but it's really an illusion created by the electric force. What

Experiments in the 1960s and 1970s showed that, just as atoms are not simple and fundamental, so protons and neutrons are made of much smaller particles that were named quarks.

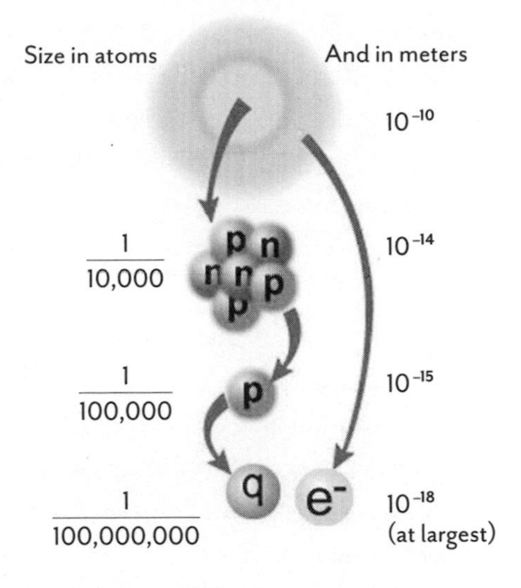

Figure 4.3. Forty years ago, high-energy accelerators revealed that protons and neutrons were not fundamental but were composed of fractionally charged particles called "quarks." Electrons and quarks are fundamental.

we call "stuff" is really empty space. Buddhists might well be satisfied that physicists too were deconstructing reality.

Then there's antistuff. The British physicist Paul Dirac was one of the pioneers of the "new physics" in the early twentieth century. By following the logic of equations that describe quantum states of subatomic particles, he predicted a "mirror" form of matter in which all the properties are opposite. Electrons should have counterparts with opposite spin and charge, called *positrons*, protons should have counterparts with opposite spin and charge, called *antiprotons*, and so on. It was amazing even to physicists when antimatter was created in the lab a decade later. Recently, it has proved possible to create antiatoms and antimolecules. Perhaps nature has managed to create antiplanets and antistars and antigalaxies? At the time, nobody knew how to look for them, and antimatter on Earth was extremely rare and difficult to make.

Returning to our exercise in cutting bread dough enough times to isolate an atom, the same might apply to protons and neutrons. If we cut an atom, would we end with protons and neutrons as the fundamental parts that cannot be cut any smaller? The monks ponder this; then

I explain. When quantum theory was first developed, nobody could break a nuclear particle into pieces and study them. Today, we speak of "smashing" atoms, suggesting still smaller components beyond protons and neutrons. In the 1960s, atom smashers in the United States and Europe produced high-speed collisions of protons and neutrons, and for a brief instant they showed hints of constituent particles, which we now call *quarks*. Quarks combine in twos and threes to form protons and neutrons and most other subatomic particles.

I've brought a box of bungees to class, and these strong elastic ropes allow us to act out the elastic forces that hold together the substructure of normal matter. Physicists discovered three families of quarks with increasing masses, six in total and called the types *flavors*, with the whimsical names *up*, *down*, *strange*, *charm*, *top*, and *bottom*. The standard machinery used to investigate atoms is the particle accelerator, and not until 1995 did an accelerator produce the last and most massive of the quarks. Quarks bind tightly to each other by exchanging gluons, which come in three types, called *colors*, named *red*, *blue*, and *yellow*. To physicists, colors and flavors are convenient labels rather than literal properties.

Every quark has a corresponding antiquark with an opposite set of quantum properties. The "glue" that keeps quarks from roaming freely and keeps protons and neutrons within the fortresslike confines of the atomic nucleus is called the "strong nuclear force" in physics. It's one of only four fundamental forces in nature. A second nuclear force is 10 trillion times weaker than the strong force. The "weak nuclear force" comes into play when atoms are very massive; it makes them decay by radioactivity. These two forces were only understood with the help of atom smashers the size of a small city. Another fundamental force is gravity, but it reveals its power only at the largest scales of the universe. For example, quarks have mass and feel gravity, but gravity is 100 trillion trillion trillion times weaker than the strong force. The fourth fundamental force is very familiar: electromagnetism. We'll return to that later.

With bungees in hand, we're ready to look at how the strongest of these forces works. The rules for quarks combining are written on the whiteboard. I get things started by turning some of us into various kinds of quarks with sticky labels. Gelek will wear a "strange" label, and I stick an "anticharm" label on my chest (neither characteristic is particularly appropriate, but then we're acting). Tied to each other by a bungee cord we form what's called a *D meson*. With bungees around their waists and bungees connecting them in twos or threes, the monks hitch up and make particles that are massive (two ups and a down quark to make a proton) and less massive (an up and an antidown quark to make a subatomic particle called a *pion*). Everyone careens around the room merrily, but fun may be obscuring understanding. We end the experiment and reset the room into a more decorous arrangement.

There are some bewildered looks, so I recap. Fundamental particles such as quarks are not really particles because they have no hard edges and their positions can't be well-defined. Their positions are described by probabilities in quantum theory. They do carry an electric charge. But when we say they have "spin," that's a metaphor; they don't actually spin like a top. The metaphors continue. Quarks come in six types, called flavors, but there's no taste involved. Physicists identify gluons by color, although they're smaller than waves of light so can't have true color. We also talk about "families" of particles, but they are no really lineal families, just similar groups.

The monks feel better now. Scientific metaphors can be confusing, but I remind them of the Robert Frost aphorism. Professional physicists use analogies and metaphors, too, often without thinking. Wave, particle, field, force, mass, time—these are all metaphors at some level. Physicists work through the math, but the metaphors suffuse and color their thinking as convenient ways to describe the ineffable. Buddhism leans heavily on metaphors, too, when it talks about reincarnation and enlightenment and suffering. Still, metaphors don't always work outside a carefully prepared context, so I know the confusion the monks are feeling. In the classroom I'm making heavy use of implicit double quotes as I talk about strange phenomena with everyday words, and you can see many double quotes in this written account as well.

Beyond the difficulty of describing situations remote from everyday experience, physicists struggle with the inelegance of the current scheme of elementary particles. There are dozens of particles, arranged in different groups, and the masses seem arbitrary rather than emerging as predictions from a theory of matter. There are also missing pieces of the picture, such as the particle that explains dark matter observations in astronomy. Dark matter outweighs normal matter by a factor of six, which makes it awkward that dark matter particles have never been detected.

If this weren't enough, there are also conceptual problems associated with the term "particle" itself. Current theory says that if an electron or a quark is fundamental, it has no physical extent and so must have infinite mass and charge density. But if it does have size, the Buddhist philosopher Nagarjuna pointed out a logical flaw: it must occupy a nonzero space, and nonzero space is necessarily composite, being divisible into smaller nonzero spaces.

If not a particle, then what? Now some physicists are speaking of strings of energy. In *string theory*, particles are byproducts or secondary phenomena. The ultimate entities are one-dimensional strings of pure energy, thirty-three orders of magnitude smaller than atoms. The mathematics only work out if there are up to six "hidden" spatial dimensions, invisible to us because they're compressed to an infinitesimally small scale. As science, string theory is below the radar of real

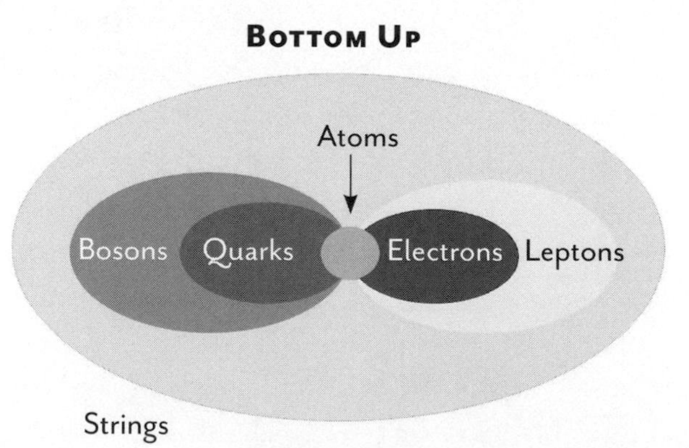

Figure 4.4. A modern or "bottom-up" view of structure is based on the hypothesis of tiny, fundamental entities called "strings," which make up the different families of particles. Strings have not yet been observed.

experience or observation, and as such it can have the feel of metaphysics, much in the way that Buddhism talks about the elusive textures of ultimate reality. Only a hundred or so physicists are technically astute enough to work with string theory. To practitioners, the theory is bewitchingly beautiful. But is it true? At the moment there's no way to know. The best testing ground is, or was, the big bang, the time at which the strings could have revealed themselves.

I'm convinced of the power of the scientific method to increase our understanding of the natural world. Yet even I get uncomfortable when my colleagues make strong assertions about strings or the cause of the big bang. Science has core assumptions that are rarely questioned and difficult to test. For example, scientists assume the natural world is comprehensible. We also assume it's subject to causality and underpinned by mathematical rationality. But without experimental verification, physics shades into metaphysics. As a scientist I should be a skeptic, and I owe it to the monks to encourage them to be skeptical learners.

Restless Energy

At breakfast, Lhakdor drops by. He's been busy up at the Library, so we haven't seen him for a few days. As always, he's full of anecdotes and stories. A young monk is impatient with his teacher. Why should we study so hard when we can go on a picnic? Don't worry about picnics, says the teacher, you should study hard and be serious. The student keeps asking but always gets the same answer. Then another monk in the class takes ill and dies. As he's being taken away, the young monk doesn't seem upset or sad, so the teacher asks why. "He's having his picnic," says the young monk.

Our response to the punch line is a mix of guffaws and groans, so Lhakdor wheels around and offers another. "A Tibetan has had a driving accident and he's telling a friend about it. 'What happened?' asks the friend. 'I was on my motorcycle at night and I saw a single light heading toward me. So I went a bit to the left to pass, but it was a truck with one light out.' His friend shakes his head and leaves. A few months later, it's the friend who is in hospital. It's another accident, another terrible collision. 'What happened?' asks the first Tibetan. 'I was on my motorcycle at night and I saw two lights coming toward me so I tried to go between the two motorcycles.'"

Rapid fire, the jokes keep coming. "A bird is flying and lays an egg, but it doesn't hit the ground. Why?" We try to guess. The bird is flying on its back? The eggs lands on water? Another bird catches it? No. He looks triumphant. "The bird is wearing underpants."

Some of his jokes are corny. Some are inscrutable. Some only the Tibetans laugh at. Lhakdor is also an endless font of aphorisms. "Life begins and ends in the quiet light of the mind." "Impossible is a word only in the dictionary of fools." "If someone breaks a leg or is crushed to death when a new temple opens, it's always auspicious." And so on. His poker face is so good sometimes we miss the joke. He has a deep and warm humanity but also a muscular view of the world and Tibet's need to be strong in the face of Chinese indifference or aggression.

Lhakdor talks about a mental exercise the monks do to feel the reality of independent origination. They contemplate the impermanence of the body and of the everyday world. "Attachment makes us afraid of death," Lhakdor says. "One monk became so scared during the exercise that he began frantically grabbing at his robes. He had to latch on to something to make sure he was there." Lhakdor says he looks forward to the way station after death where you can pass through matter, go anywhere, and be anything you want. Suddenly he turns and looks at me intently. "I don't know you," he says, sending a chill down my spine. "I'm seeing you for the first time."

Back in class, Lhakdor has given me a jumping-off place for our continuing discussion about matter and energy. Matter is very attached to itself. It can be rearranged, but mostly it stays put. On the other hand, energy has many guises. It's as varied as a radio wave or a tank of gas or a coiled spring. Physicists encompass such diversity with a definition that smacks of Zen Buddhism: energy is anything that can cause change. But energy is also like the Hindu god Shiva, who was incorporated into the Buddhist tradition. Shiva is creator, transformer, and destroyer. In one dance, Shiva provides the energy that sustains the cosmos, and when the dance stops, this universe will end and a new one will begin. In the natural world described by science, energy is also restless and constantly changing.

I ask the monks to count the energy transformations when they talk to a friend on a cell phone. It takes a while to tease out all of the steps: electrical and chemical energy from the brain is turned into mechanical energy of vibrating vocal cords, driving a compression wave of sound that travels through air, which moves coil in the microphone of the cell phone, which is converted into an electric current, which is emitted as a radio wave into the atmosphere, and after some boosting by various cell towers it follows the changes in reverse, ending as the electrical and chemical brain activity of the person receiving the call.

Our second example is a monk eating a momo, the Tibetan dumpling. The flour in the wrapping and the vegetables inside have turned radiant energy from the Sun into chemical energy. The monk will store and later utilize that chemical energy to walk to the stall to buy more momos, thereby continuing the cycle. Buddhist monasteries are vegetarian as a nod to sentient creatures and for health reasons. But the no-meat rule is mild, and if the monk is slightly naughty and sneaks off-site to eat a mutton momo, there's an extra step in the energy chain. That's because the sheep converted the chemical energy stored in grass into the chemical energy stored in its muscles.

The dizzying transformation of energy between easily recognizable and barely recognizable forms sits above an inviolable law of physics: the conservation of energy. Energy cannot be created or destroyed; it can only change forms. If the accounting of the different energy forms is done properly, this is always true. The monks are comfortable with the mutability of energy and the interplay of its diverse forms because it accords with their core ideas of impermanence and interdependence.

During a break I slyly place a "nodding duck" on the table where the projector sits. The monks come back from butter tea and biscuits and are immediately engrossed. The small toy duck rocks back and forth, dipping its felt beak into a glass of water, oblivious to all the attention. How does it work? Where does the energy come from? Will it ever stop? We work through the possibilities. Thupten raises his hand: the felt beak evaporates water so is always slightly cooler than the body. Yes! Thupten doesn't nail it but he's on the right track.

The duck is a heat engine. Here's how it works. Water evaporates from the felt on the duck's head. That lowers the temperature of the head. The lower temperature causes some of the vapor in the hollow head to condense. The cooling and condensation cause the pressure to drop in the head. Higher pressure in the warmer base pushes liquid up the neck. As the liquid rises, the bird becomes top heavy and tips over. When the bird tips over, the bottom of the neck tube rises above the surface of the liquid in its body. This lets a bubble of warm vapor rise up the tube. Rising vapor displaces liquid, which flows into the bulb at the bottom. That weight makes the bird return to the vertical position. And the cycle repeats. The toy uses half a watt of evaporative heat flux to drive its motion. The source of the energy is a temperature gradient between head and body. The duck isn't thumbing its beak at logic or reality; it's demonstrating one of the core principles of physics.

I've brought a small stack of stickers with the NASA logo to India, and I've been waiting for the right opportunity to reward special contributions in class. I bestow our first on Thupten. My monks are too advanced in their Buddhist training to be envious, or attached to a mere decal, but their eyes do open wide as Thupten plasters the iconic red, white, and blue logo on his burgundy robe and (perhaps I imagine this) his chest swells perceptibly.

Now I hand out three different sheets of thermoplastic film, which responds to heat by turning a color from red to blue depending on temperature. It's another demonstration of the conservation of energy, as invisible heat converts through a chemical reaction into visible light. Each sheet has a different range of sensitivity, and the monks attempt to figure out which of them can register cool, warm, and hot. They're creative in this activity, holding the sheets against their foreheads and arms, breathing on them, and placing them over the air conditioning vent and over the fan grille on my laptop computer. They're working like true scientists. Jigme and Dawa are the class clowns again, making ghostly face prints on the sheets that go from blue to red before fading to black.

Forms of Energy

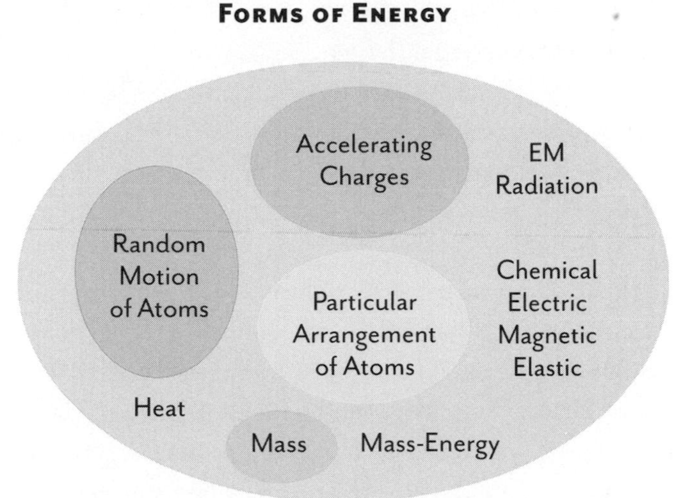

Figure 4.5. Microscopic forms of energy are diverse and include heat or random motion, radiation from accelerated motion, energy stored in chemical bonds, and energy stored in the form of mass.

Now we make a connection with the behavior of the universe. As a hot plate on a stove cools down, its color changes from yellow, to orange, to dull red. That is because the wavelength it emits is growing longer. After the color fades from sight, the plate is still pumping out radiation, just at a wavelength too long for our eyes to detect. When the universe was an infant it glowed bright in gamma rays. After hundreds of thousands of years it cooled enough to emit light. Now it's a thousand times bigger and cooler. The radiation waves have been stretched, moved through the red zone (that is, redshifted), and become microwaves. Except, as I remind the monks, we could see the microwaves as interference on the old television.

The shifting interdependence of energy forms is easy to demonstrate. Two common ones are kinetic energy, the energy of motion, and potential energy, the latent energy of an object subject to gravity. I have a tennis ball with me in class. I roll it between the rows of mats and the monks all turn to watch it pass. I've converted chemical energy in my muscles into kinetic energy of the ball. Next I hold the tennis ball up and drop it. This time I didn't exert myself but the ball still ended up with kinetic energy from the Earth's gravity field. Gravitational potential energy was converted into kinetic energy.

Time to make it a bit more interesting. Soon after arriving, I started scouring the monastery and surrounding area for items that might be useful in the classroom. On an early evening run I spotted Indian kids in a nearby village playing soccer with a threadbare ball. I asked them to be there at the same time the next day and I returned with enough rupees for them to buy two or three brand-new balls. Skeptically, they pocketed the cash and handed me the ball, with its webbing showing through the leather almost everywhere—these Westerners are clearly crazy. Later that day, in my room, I cut into it, filled it with sand, and stitched it up, then attached a hook between the leather panels. Now, during our afternoon break, I suspend the ball from the ceiling of the classroom using an existing hook. It's heavier than a bowling ball and I worry the hook might come out so I test it with my weight. The ball hangs down almost to the floor, a drop of ten feet.

The monks are watching intently. I pull the ball up along its circular arc and stand on a chair, positioning myself so that the rope is taut and the ball is barely an inch in front of my nose. Long ago, I saw this trick done by a physics professor in a large auditorium using a bowling ball, and I recall the importance of not leaning forward inadvertently when the ball is released. That's the way to break my nose. Standing straight and tall, I release the ball. Thirty-two monks inhale as one as the ball sails with a whoosh toward the floor. Their eyes follow it as it swings to head-height on the far side of the room, then rushes back toward me. I wince, but the massive ball stops an inch in front of my nose. I laugh with relief. "Physics is pretty reliable," I say. "The energy has been conserved."

We've just watched the smooth exchange of kinetic and potential energy. Kinetic energy is maximum when the ball is closest to the ground and moving quickly, but drops to zero as it stops for an instant at the end of arc. The total energy is the sum of the two and is constant. The ball won't swing forever, of course, because gradually energy is lost by the friction at the hook and from resistance with the air in the room. We gather around my computer and I show an animated example of the same thing in astronomy: an elliptical orbit. As the Earth or any planet or a comet gets closer to the Sun, it speeds up, and as it gets further away, it slows down, a variation described mathematically by the laws of Johannes Kepler, a German astronomer working at the turn of the seventeenth century. There's no friction in the vacuum of outer space. So the sum of two types of energy is fixed and perfectly conserved. Whether in the swing of a pendulum or the swinging motion of an orbit, we're watching the precise mechanism of Isaac Newton's clockwork universe.

Energy conservation works across the universe. Galaxies zoom away from each other, but they also tug on each other; kinetic energy and gravitational energy strike their balance. The two trade off as the universe grows. Slowing expansion means less kinetic energy of motion, but the gravitational potential energy grows as the universe gets larger.

"Now, where's that tennis ball?" It's handed to me and I toss it up a few feet. It falls back into my hand. We're seeing the conversion of energy of motion into gravity energy as the ball rises and then back again as it falls. What if I go outside and toss it higher and higher? Several monks offer answers. That's more energy for the ball, says one. But it will still come back down, another says. It might leave the Earth forever, says a third.

"Exactly!" I'm using exclamations more often with the monks, and they know what it implies in English. I think I've borrowed this dramatic flourish from Geshe Lhakdor, who uses exclamations frequently for emphasis, often riding over the top of my comment even as he agrees with it. I smile to myself. Perhaps I'm a Lhakdor wannabe.

If it has enough energy of motion, the ball will keep slowing down but nevertheless leave the Earth's gravity. It will never return. The universe acts the same way. If the energy in the expansion is sufficient, it will never be overcome by gravity. The universe will expand forever. That's what cosmologists called an "open universe," I explain. If the energy in the expansion is less than a certain amount, galaxies tugging on each other by the force of gravity will eventually overcome the expansion and the universe will collapse. That's a "closed universe."

Astronomers have measured the current expansion rate and also the

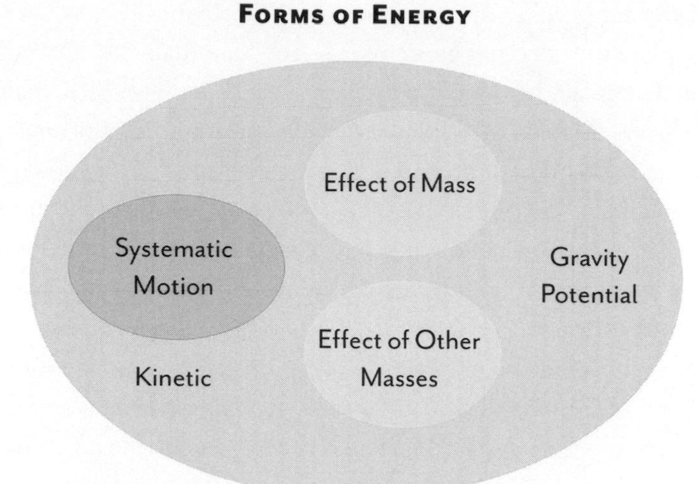

Figure 4.6. Macroscopic forms of energy include gravitational energy and energy of motion, or kinetic energy. An orbit or a pendulum swing has a continual exchange between gravitational and kinetic energy.

past expansion rate by looking out in space to times long ago. They think that the universe has slowed down since the big bang, caused by the gravitational braking of normal matter and also the more abundant dark matter. Yet the brake hasn't been enough to overcome the expansion. Hold that thought, however, because the push-and-pull gets even weirder. Having slowed after the big bang, the expansion has picked up speed in the past 5 billion years. When astronomers detected this, they attributed it to a mysterious entity we call "dark energy." If dark matter is an enigma, dark energy is a cipher. But it puts an exclamation mark on predictions of the cosmic expansion. The universe will expand forever!

○○○○○

On our second weekend in India, Paul and I spread our wings. Saturday finds us at the Kangra fort, the largest in the Himalaya, first mentioned in Alexander the Great's war records. The huge fortress sits on top of a towering cliff at the junction of two rivers, monolithic and impregnable. It's brutally hot, and the air parches our throats. We have the place to ourselves, along with a few monkeys.

Paul swore he'd never ride in a three-wheeled auto-rickshaw, but there are no taxicabs at the fort gate when we want to leave, so we bundle into one of the flimsy conveyances. Powered by lawnmower engines and riding on wheels the size of a spread hand, they're an essential form of transport in India and all developing countries. We feel like we're on Mr. Toad's Wild Ride as we barrel through the towns of the Kangra valley, swerving between brightly painted trucks, and wincing as animals laden with wood and women laden with rice magically melt to the roadside as we whip by.

Bryce says we're invited back to the Tibetan Children's Village in the town of Suja to stay with a man named Nyima, who is a science teacher in the town school (and not to be confused with one of our monks, Geshe Nyima). Tenzin drives us over to the town of Suja, which is higher on the slopes of the Himalayan range and therefore cooler. The Tibetan Children's Village is a self-contained campus on the outskirts of Suja. The monsoon has left the ground sodden and glistening.

When we meet Nyima he's apologetic because he's sick. It's more than a cold, probably something respiratory. He's clearly overworked. Nyima is about forty. As a senior and respected science teacher, he puts in fourteen hours a day and is paid five thousand rupees a month, about fourteen hundred dollars a year. That's less than some of the kids get that we've seen begging in McLeod Ganj. His wife is a nurse at the school clinic. She worked fifteen hours and saw 130 patients the day before we arrived. They have a son and a daughter.

Nyima wants to be the perfect host, so he treats us to entertaining accounts of life in his region. He tells us that nearby is one of the world's best paragliding sites; it gets wonderful updrafts at the front edge of the tall mountains to the northwest. A few weeks ago, two Russians got carried high above the mountains and were never seen again. But they'd been drinking vodka before their flight, Nyima says, and he laughs. I'm shocked by the story, but make a mental note—paragliding would be very cool to try.

After a light lunch Paul and I have a walking tour of the "village," which is really a kind of residential school. There are a number of Tibetan Children's Villages on the India border, built to handle the influx of refugees, and they operate like boarding schools. This Children's Village has nearly nine hundred children and just sixty teachers and staff. The regimen is tight. The prayer bell rings in the morning at five-thirty. Breakfast is at six fifteen, and classes start at seven. No cell phones are allowed. There's a single computer lab with fifteen cast-off machines. Kids can't get furloughs or leave the village except to visit family. Some have families nearby who are too poor to house them. But most are like orphans, since their families are back in Tibet or widely scattered across India and unable to travel.

It sounds grim and dispiriting, and I'm sure there must be pockets of melancholy and homesickness, but the overall mood is upbeat. Smiles and waves greet us everywhere. The children are inventive in their play, as they must be when there's no equipment. Along the length of the dirt soccer pitch, we see a wall painted with letters ten feet high: "Others before Self."

We spend the afternoon with Nyima's son Ugyen. Ugyen is a pistol. He's thirteen, his English is excellent, and he wants to be the first Tibetan astronaut. He shows us his astronomy scrapbooks. They're filled with articles on topics from antimatter to asteroids, laminated and annotated in English with his precise, spidery script. He knows a ton of astronomy and follows the space program keenly.

Ugyen leads us several miles through rice paddies and pine forests to the Chokling Gompa monastery. We remove our shoes and stand at the back of the prayer hall, where several hundred monks are gathered to celebrate the end of a major teaching. The prayer hall is sumptuously decorated with banners and tapestries with a row of ten-foot-high butter sculptures up front. The monastery is home to Neten Chokling Rinpoche, a man tied to the oldest tradition of Tibetan Buddhism. His lineage as a reincarnated lama goes back to the eighth century. While waiting for the rain to let up so that we can start walking back, the Rinpcohe emerges and is driven off in a small SUV. This imposing and handsome man has acted in and directed movies; many American women have been smitten by his good looks. He's a Buddhist rock star.

We walk back with furled umbrellas under lowering skies. After fording one swollen stream, we stop to remove leeches from our feet. Dinner at Nyima's home is a feast of momos, mutton, chicken, okra,

Dharamsala is a hill station in Himachal Pradesh, that is famed for its large Tibetan community centered around the activities of the Dalai Lama. Tibetans hang prayer flags at ceremonial locations around the town.

An acrobatic display of agility and steadiness being performed by local residents of Dharamsala.

The Himalayas are often shrouded in clouds or mist, while the lush Kangra Valley below is a place where local Indians grow rice, using methods that haven't changed for centuries.

Local markets sell vegetables, fruits, and flowers, creating a feast for all the senses.

Dharamsala is the center of the Tibetan exile world in India. Buddhist monks engage in prayers daily, with a first session that often starts before dawn.

Religious ceremonies are accompanied by chanting and music played on copper trumpets, cymbals, gongs, and bells.

Following the 1959 Tibetan uprising there was an influx of Tibetan refugees who followed the 14th Dalai Lama into exile. His presence and the Tibetan population has made Dharamsala a popular destination for tourists.

Tibetan monasteries typically have a central courtyard when monks gather for debates. These animated exchanges can continue long into the night.

Debating is an important tool for monks to learn Buddhist traditions and to question and refine their knowledge.

Dawa Dorjee

Dhondup Gyaltse

Gedun Gyatso

Gelek Gyaltsen

Jime Gyatso

Kalsang Gyaltsen

Konchok Choephel

Chundrup Gyatso

Lobsang Choephel

Lobsang Sopa Namgyal

Lodeo Sangpo

Ngawang Gyatso

Ngawang Lobsang

Ngawang Norbu

Ngawang Nyingtop

Ngawang Sherab

Nyima Tashi

Sherab Tenzin

Sonam Choephel

Sonam Wangchuk

Tashi Phuntsok

Tenpa Phakchok

Tenzin Choegal

Thabke Lodoe

Thupten Kunkhen

Thupten Nyinghe

Thupten Tsering

Tsering Phuntsok

Yeshi Choephel

Yundrung Konchok

Participants in the Science for Monks program. Teachers included are Gail Burd (back row), and Richard Sterling, Chris Impey, and Mark St. John (middle row).

The author with the translators. From left to right: Tenzin Paldon, Sangey Tenzon, Chris Impey, Tenzin Sonam, and Karma.Thupten.

The author with Geshe Lhakdor, Director of the Library of Tibetan Works and Archives.

Bryce Johnson, the workshop organizer, with the author's son, Paul.

The classroom for the Science for Monks workshop. It's a low-tech environment but often filled with light and lightness.

Karma helps Sonam Wangchuk experiment to model experimental errors.

A classroom of Monks is both a serious and a solemn place.

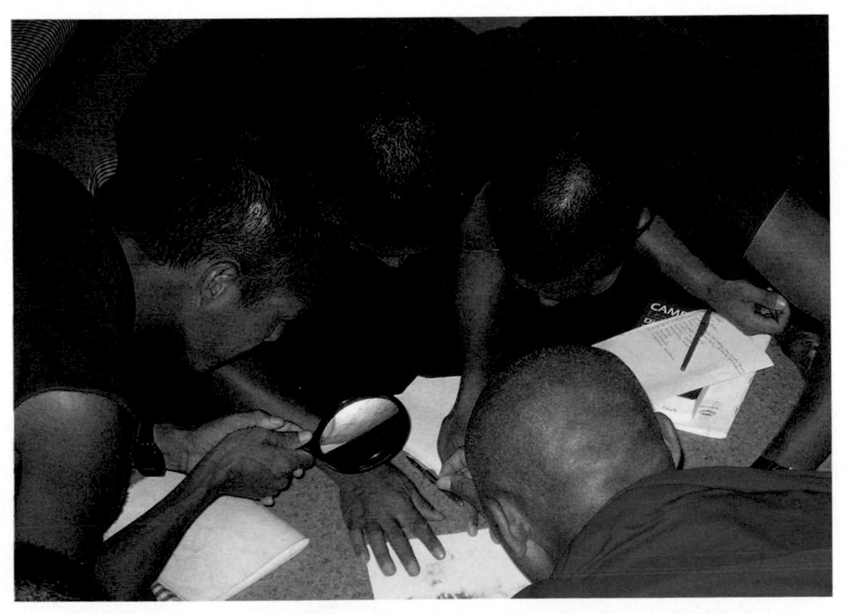

Grains of sand are used to estimate the contents of space and the number of galaxies in the universe.

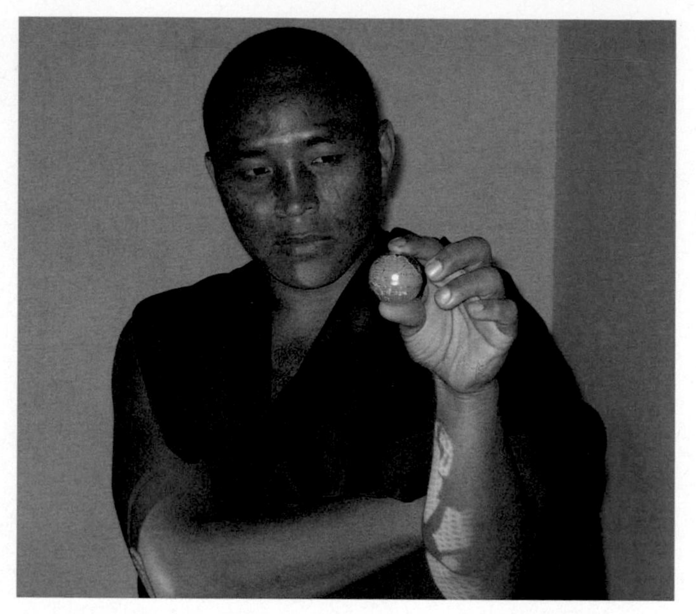

Thupten Tsering B holds Earth in a scale model of the Solar System. Other monks leave the classroom to take up the positions of different planets.

ABCD cards are used for voting and as a method for answering questions.

The distance between the dots being drawn on an expanding balloon trace the Hubble law that describes the expanding universe.

After an activity on cosmic expansion, a light hearted end to the universe!

A monk is a force of nature. Geshe Lhakdor attempts to frighten a visitor.

Zoob is being used to model the forces of nature. The toy serves to model hypothetical as well as actual physics.

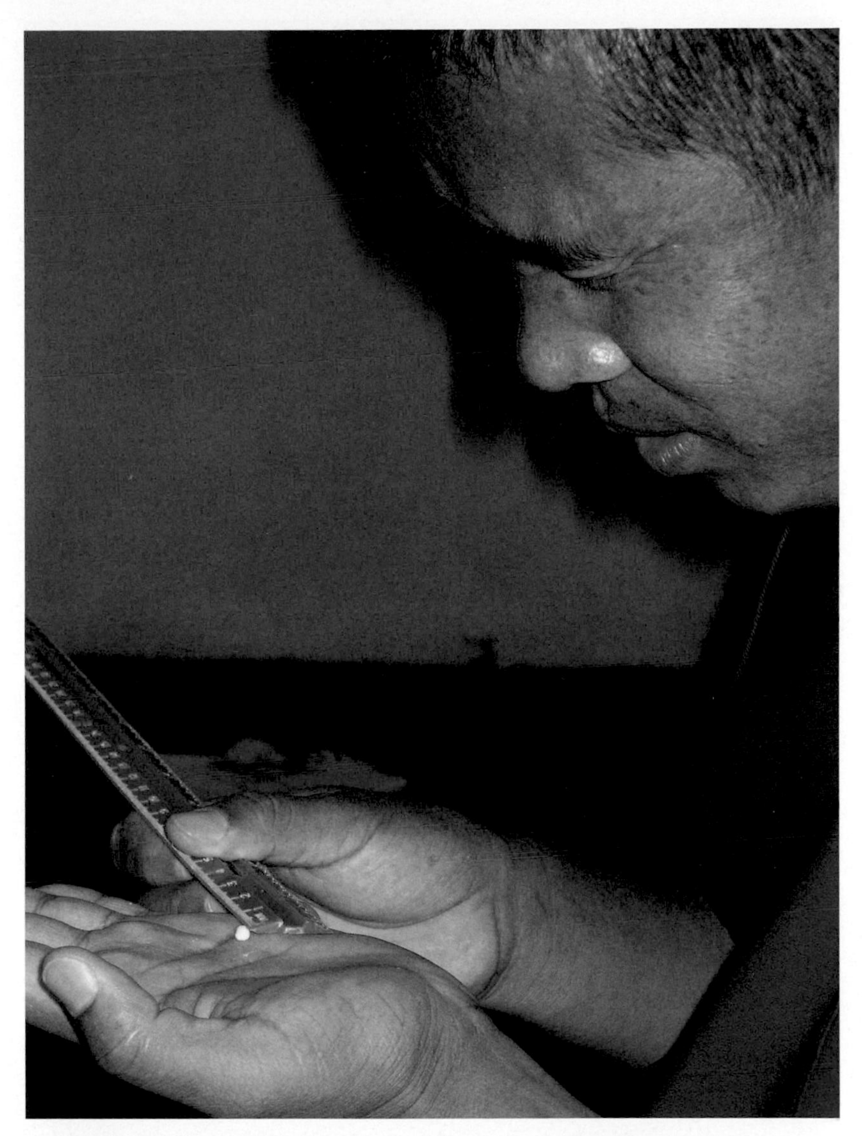

Gelek Gyaltsen subdivides a ball of dough. Starting with a lump the size of a soccer ball, he extrapolates the process to atomic scales.

Masked monks prepare to act out the early big bang from the blinding heat of the singularity through the "freezing out" of the four forces.

Ten thousand poker chips stand in for galaxies in a simulation of the formation of structure in the universe.

Albert Einstein, the Geshe of time and space, pays a visit to the classroom.

Kalsang Gyaltsen and Gelek Gyaltsen debating a cosmic timeline.

A celebration of the unity of life at a hilltop holy site.

Geshe Nyima Tashi sports a tattoo of the spiral galaxy M51, that is also called the "whirlpool."

The monks tether ideas and challenge themselves using formal debates.

Debates have ritualized rules and are very competitive, but the monks never get attached to the outcome.

Plastic flowers are used to show the evidence that's required to distinguish living from inert matter.

Our lives are connected since the big bang by the story of stardust.

Despite eighteen hour days filled with study and lessons, there is time for recreation. The form of shuffleboard called karom is particularly popular.

Classes occupy six days of the week. On a day off, the group hikes through a pine forest and pauses for a rest.

The monks displaying their certificates for the successful completion of the Science for Monks workshop.

potatoes, cauliflower, and tsampa. There are many little side salads made of ingredients I can't decipher. Nyima has no spare room, so we sleep under heavy woolen blankets in his shrine room, with candles flickering on the mantel and His Holiness smiling down at us beatifically.

Making our way back to the College for Higher Tibetan Studies the next day, we take one of Bryce's recommendations. At the nearby village of Ahju, there's a narrow-gauge train, and Bryce said it traverses the Himalayan foothills and can carry us home. We walk down to the tiny station house. From behind the grille, a slender Indian man sells us second-class tickets. They cost ten rupees each, or fifteen cents. Not bad for a three-hour train ride. To my surprise he gestures for us to come around the back and into the station house.

The room is sparsely furnished and decorated with iron and brass fittings that probably date back to the construction of this branch line in the 1920s. The walls have holders for memos with archaic titles like, "Report of Most Serious Incidents," and "Problems Arising from the Issuance of Tickets." The stationmaster is tall, debonair, and dark-skinned, with sleepy eyes and graceful fingers. He wears pale blue pajamas with fine yellow stripes.

The stationmaster from Kashmir is obviously unhappy with his posting to this toy train in a tiny village, far from his family. He's educated and well-read and starved of good conversation. We talk easily about science and the latest discoveries. He's well informed about everything from genetics to cosmology. When the subject turns to Indian mathematicians, he lets slip that he has many notebooks on the subject. "They are mere scribblings, probably just doodles, but perhaps you could come to my house and look them over?" Paul looks nervously at me. I politely make our excuses and we head down to the platform.

As we board the train, I wonder, have I turned my back on the next Ramanujan? Srinivasa Rumanajan was the Indian autodidact who emerged from a small Indian village to make extraordinary contributions to number theory and mathematical analysis. He got a PhD from Cambridge but died from tuberculosis at age thirty-two.

The train is fun but exhausting. We ride from hill towns down through clefts in the forested slopes and eventually emerge in the fields and paddies. One stop stretches to two hours when they have to swap the engine. A family of five sits across from us. The parents speak almost no English but offer us food, which we reluctantly decline. The three brothers are seven, ten, and thirteen, and they befriend Paul, "borrowing" his iPod and peppering him with questions. They're from a small town, and when I comment on their perfect English, they're surprised at my surprise: "Of course we speak good English!" They are three reasons that India will be a force in the twenty-first century. I'm inspired by their optimism and promise.

Relatively Speaking

The Himalayan train ride is actually good preparation for our next class, for train and trams have a significant role in the history of physics. The pioneers of quantum theory crisscrossed Europe to visit each other on trains, and often talked physics on long train trips to vacation together in the Alps or the forests of Bavaria and Denmark. Albert Einstein's daily ride to work at a patent shop in Bern, Switzerland, helped him grapple with his boyhood question, "What would the world look like if you rode on a beam of light?" He used the journey to pose deep questions of time and motion. Einstein's tram travel helped him arrived at his theory of relativity.

At the time, it was known that light was just a sliver of a broad spectrum of electromagnetic radiation. The English scientist James Clerk Maxwell had developed elegant mathematical formulas to describe how all these types of radiation traveled at the same prodigious speed: three hundred thousand kilometers per second. In its early form, Maxwell's theory seemed to suggest that even when things are in motion, the speed of light is the same. But how could that be? Surely if the source of light is moving relative to the recipient, that motion adds or subtracts from the speed of light. As Einstein sat on his daily tram ride, he put the question in a different way: Can you catch up with a photon, or evade it by moving fast enough?

To test this I get ten volunteers among the monks. With some difficulty I've borrowed five flashlights, or torches as the Indians call them. A few are feeble but they all give off some light. We clear out half the room for the experiment. Five monks stand along the far wall of the room. I ask them to measure the speed of the photons when they arrive. They're to assume they have equipment that can do the job with precision. (I don't actually give them any equipment; this is a thought experiment and they seem okay with that.) The five other volunteers line up in the middle of the room and each gets a flashlight but each is given different instructions.

One shines his light at a monk on the opposite wall. That's Experiment A. The second shines his light at another monk on the opposite wall as he's walking toward him. That's Experiment B. The third shines his light at a monk on the opposite wall while running toward him. That's Experiment C. The monks doing Experiments D and E have slightly more difficult jobs. They have to shine, walk, and run away from the monks on the opposite wall, respectively, while shining their lights back at them over their shoulders. I write the speed of light on the whiteboard as three hundred million meters per second. This is the constant labeled "c." I say we can assume walking speed as one meter per second and running speed as two meters per second. Before starting the experiment, I hand the five "receiving" monks an envelope with the correct "answer" written on a piece of paper inside.

As always, the monks relish the physical part of the activity, walking and running with gusto. When the action is over, I bring the rest of the monks to the whiteboard and ask them to estimate the speed at which the light will reach the receiving monks in the five different experimental setups. There's some disagreement, but after vigorous debating they arrive at a consensus. The arithmetic seems simple. Experiment A should give 300 million meters per second, and with the other four the source motion either adds or subtracts from light speed, so they predict 300,000,001 for Experiment B; 300,000,002 for Experiment C; 299,999,999 for Experiment D; and 299,999,998 for Experiment E.

"Are you sure?" I ask. Smiles and nods all around say that they are. We turn to the receiving monks, and ask them to open their envelopes. There's great anticipation, just as there is at the Academy Awards.

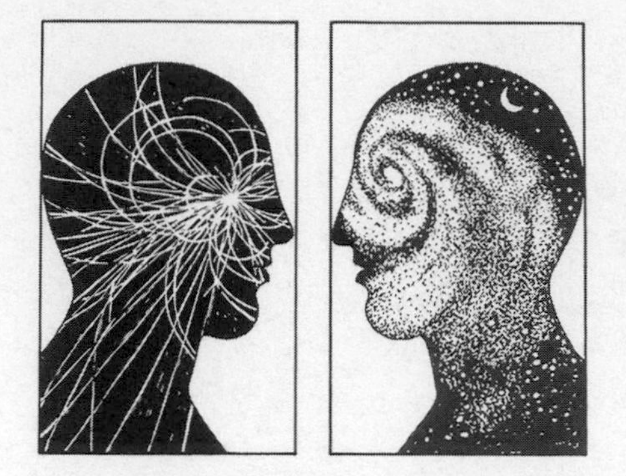

We are made of tiny particles and we are part of an enormous galaxy, yet we keep both within our heads.

Figure 4.7. Ancient Greeks showed that the nature of the very large and the very small could be explored without equipment. Einstein also used thought experiments to make deductions about nature.

Each one says 300 million. This was the result of a real experiment conducted by physicists, I tell them. The monks are visibly taken aback.

The actual measurement, done at the turn of the twentieth century, was even more decisive. As background, a century ago things like water and sound were a model for how all waves moved in the universe, including light. Everything needed a medium in which to travel. Therefore, it was assumed that *something* had to be waving or oscillating when light moved, even in a vacuum. Physicists referred to this invisible substance as the *ether*. To detect the ether, they looked for a change in the arrival time of light, in the form of photons, as the Earth orbited the Sun. But the arrival time of the photons never changed. Regardless of the Earth's motion, photons always have the same speed, three hundred thousand kilometers per second. There's no ether.

Einstein elevated this empirical fact to a postulate of his special theory of relativity. He labeled it "special" because it only applies to the special case of uniform relative motion; for the case of accelerated motion, which includes gravity, he invented the "general" theory of relativity. If the speed of light really is a universal constant, there are some very strange implications. Any object moving close to the speed of light shrinks in the direction of motion. Time also slows down for any object moving close to the speed of light. These effects aren't tricks or illusions; they're actual physical effects that have been observed routinely in physics labs for a century. Speed is distance divided by time, so the shrinking of space and the stretching of time for fast-moving objects act to preserve the constancy of c. In addition, the mass of any object increases as its speed approaches the speed of light, such that the object can never move faster than three hundred thousand kilometers per second.

Relativity is "below the radar" in our everyday world since even bullets and fast rockets move sluggishly compared to light. However, for fast-moving fundamental particles, all the weird distortions of space, time, and mass are seen. As bizarre as they sound, the distortions are not just theoretical constructs; they're real.

Stuff and light seem so different, I remind the monks. Matter has heft and substance and resists changes to its motion; radiation is intangible and evanescent and fleet-footed. Despite this stark difference, Einstein formalized a profound connection between the two. I take a fat black marker and write his famous equation on the board, $E = mc^2$. Many of the monks nod in recognition. It's a superstar equation that people quote even when they don't know what it means. From his musings on the speed of light and deep thoughts during his train rides, Einstein came up with a fundamental definition of physical reality: matter and energy are the same thing, and thus interchangeable. He called this "the most important upshot of the special theory of relativity."

The iconic equation implies that energy and matter are equivalent. Viewed in one way, since the square of the speed of light (c^2) is a very large number, the mass equivalent to a large amount of radiation is very small. Photons have energy, so they also have mass, but it's infinitesimally small. However, looked at in the other direction, Einstein's theory becomes astounding. It says that a tiny amount of matter, such as an atom, embodies a huge amount of energy. Einstein said that mass is like "frozen energy." When it's released or harnessed, as in nuclear power plants and atomic bombs, a single atom's energy is truly impressive. With Einstein's equation the conceptual wall between stuff and radiation is dissolved.

Getting energy from stuff preoccupies humans. We eat to stay alive. We burn wood and oil and coal to heat and light our homes. We use those same energy sources to build buildings, cars, and computers. Most of this energy still comes from rearranging molecular bonds in a chemical process. Living mostly as vegetarians, I tell the monks, you eat formerly living things that in turn "ate" sunlight, referring to the photosynthesis, which captures sunlight and makes the bulk of a plant. I'm an omnivore, so in addition to vegetation I eat formerly living things that ate formerly living things that ate sunlight. Modern civilization operates by extracting the crushed remains of formerly living things—namely, fossil fuels—from deep underground and liberating their chemical energy. We live—literally and metaphorically—on life itself.

Now I do something slightly cruel. I uncover a small plate on the front table. There's a momo on it, still warm from the kitchen. I walk with it down the aisle and three dozen pairs of eyes pivot to follow its progress. The piquant aroma reaches them, and several monks unleash soft sighs. Mutton. It's a lamb momo that the teachers enjoy daily, not the vegetarian kind served in the monastery. By tempting them I've deviated from the Noble Eightfold Path, but I hope that it won't hurt my karma because it's in the cause of science.

A momo weighs about fifty grams and it has a nutritional value of 150 calories or 600,000 Joules, in the energy units that physicists use. Half of that comes from fat, even though it's only a third of the weight, and that's why humans crave fatty meat and why vegetarians are taxed in getting their calories from other sources. In India, the average daily calorie intake is 2,300, equivalent to fifteen mutton momos, or twenty vegetarian momos. The body actually uses roughly a third of the energy content of food.

Now imagine, I tell my audience, that they could get their energy directly from sunlight, like a plant does. I grab Jigme playfully by the shoulders and get him to stand up. If you lie down, I tell him, you have about half a square meter surface area, and that would receive 500 Joules every second from the Sun. By sunbathing, and after accounting for the low efficiency of photosynthesis, you'd need an hour of lying in the sun to get the same energy as from eating one momo. You'd have to lie in the sun all day to survive, and it would leave no time for prayer or meditation. There must be a better way to get energy.

Here's what the Sun does. Due to the high pressure and temperature caused by its enormous mass, protons deep inside the Sun move so fast they overcome their mutual electrical repulsion and stick or fuse due to the nuclear force. In three violent stages, protons convert into the nuclei of helium atoms. But one helium atom has slightly less mass than four hydrogen atoms. The result is 0.7 percent less than the sum of the parts. The "lost" mass emerges as energy: sunlight. It's a vast amount of energy, about 4×10^{26} watts, corresponding to the loss by fusion of 4 trillion kilograms every second. That sounds scary, but the Sun is so massive that it will shine for billions of years even at that fantastic rate of mass-energy conversion. About 1,350 watts arrives at each square meter of the Earth's surface, enough to keep a dozen light bulbs lit.

Suppose we could harness mass-energy from momos the way the Sun does from its hydrogen. That would be 100 million times more efficient than getting chemical energy from the same mass. But wait. That's only 0.7 percent of the mass-energy released. Logically, then, to get a 100 percent release of mass-energy, you'd need a reaction that annihilates the matter completely. That's the role of what physicists call *antimatter*, which is presumed to have existed at the origin of the universe, annihilating *nearly* all the matter, yet leaving a portion behind and producing a massive amount of radiation, now seen as microwaves.

Since the origin of the universe, antimatter has become rare on Earth and elsewhere. Thankfully so, I tell the monks. It annihilates any matter it contacts and releases the pure radiant energy of gamma rays. Alas, if only we could use it without being annihilated, we could create energy with 100 percent efficiency. To meet the world's energy requirement for one year, all you'd need to do is make fifteen thousand antimomos (which is rather hard to do) and collide them with fifteen thousand momos. Compare that to the 40 trillion momos it would take to feed the world for a year with a healthy but rather monotonous diet, and the advantage of perfect mass-energy conversion becomes obvious.

For all this speculation, momos will not solve our energy crisis. So the monks and I go back to the universe and its natural energy sources.

These are the stars in our galaxy and the billions of other galaxies. A total of 10^{22} stars pump out 10^{45} watts, enough to power 1,000 billion billion billion civilizations like ours, or more if they're not as profligate.

We think of stars as giant light bulbs in the sky, but the light we see is only a minor by-product of fusion reactions. Stars are really chemical factories that leak light as a byproduct of forging elements from lighter to heavier in the periodic table, just as a factory making widgets might leak light from its windows. The ultimate source of the light is gravity.

As lunch approaches, and our stomachs growl after too much talking about momos, we consider our good fortune. Something invisible and dark, the force of gravity, has led to a cosmos filled with light and generated the elements that give rise to life and let us, in our best moments, shine.

Playing with Zoob

When we return in the afternoon, we have a lot of information to digest as well as our lunch, so we segue into discussion. The monks are challenged but excited by these concepts of matter and energy. Of particular interest is my statement that the universe is empty space, a vacuum. I eavesdrop on one of the groups and Tenzin translates for me. Almost all the monks are saying, "That really can't be true."

"Oh, yes it can," I counter, and the statement is consonant with Buddhist cosmology. Modern physics reveals that the solidity of matter is an illusion caused by the electrical forces; atoms are overwhelmingly empty. Atoms aren't objects in space; they're shifting waves of probability. Matter oscillates in and out of existence; the microscopic world is unpredictable and fluid. While much of this plays out on scales too small to see, it's also the framework for thinking about the universe, which was once small enough and hot enough to be governed by quantum processes. There is no objective reality. As Heisenberg said, "What we observe is not nature itself, but nature exposed to our method of questioning."

A similar sense of reality was conveyed by the influential Buddhist philosopher Nagarjuna, who lived in the third century BCE and who is sometimes referred to as the "second Buddha." Nagarjuna talked about *sunyata*, the Sanskrit word that means emptiness, but also implies latency and the absence of inherent existence. He also advocated the

Middle Path, which avoids the extreme of thinking that things are permanent and independent of their parts, but also avoids the extreme of nihilism where things don't exist. Naturally, with an idea this rich, our discussion rages for an hour.

Suddenly I detach from the scene. The babble of Tibetan around me ebbs, and I'm in a bubble of near silence. I watch the monks laughing and arguing and feel strangely flat. What if the middle path is simply a muddle? Immersion with Tibetan monks makes me sympathetic to many aspects of Buddhism. It seems like a very sensible religion. It advocates compassion and self-awareness. It concentrates on right behavior rather than idolatry, whether toward gods or human inventions and heroes. Nevertheless, Buddhism does have a kind of theological outlook, where the purpose of life is to lessen suffering and end the cycle of rebirth. This doesn't resonate with me, especially the idea of rebirths, not to mention the prospect of lessening the intractable fact of human misery. Perhaps an agnostic is just an atheist without the courage of their convictions. As the debate flows silently around me, I'm adrift, floating on a raft to a destination unknown.

It's been a very long day in the classroom. My energy is flagging. Tenzin is translating and he's exhausted, too. The Sun slants across the monks' faces, and fading light sparkles as it catches dust motes in the early evening air. Our energy is in short supply, even though we understand Einstein's equation a little bit better now. The only thing to do is to finish strong by piquing the monks' curiosity. It's time for Zoob.

Zoob is a toy, but a highly educational toy. To get the monks thinking out of the box about physics, Zoob just might do the trick. The Zoob I've carried to India comes in a clear plastic box the size of a small suitcase. Inside are five hundred colorful plastic pieces. It's a clever construction toy, like Lego on steroids. The hard plastic pieces are red, blue, green, yellow, and gray. Each color is a slightly different shape. The five different shapes have three attachment points: at either end is a ball or a jaw socket, and at the midpoint is a notch.

The five different pieces connect in twenty different ways and some shapes form more connections than others. I don't tell the monks any of this. Better to let them make the discoveries by themselves. That

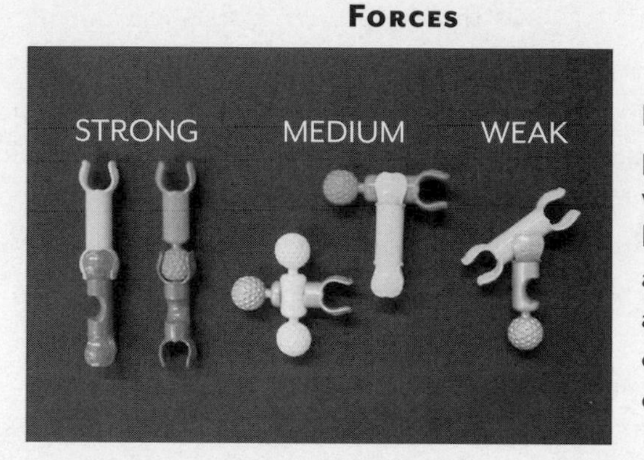

In physics, particles interact via forces.
In Zoob, there are five particles and five forces or ways of connecting.

Figure 4.8. Zoob is a toy with five shapes and five colors, which can combine to produce an essentially infinite diversity of structures. It's a physical analogy for the forces of nature and formation of structure.

was the intention of Michael Joaquin Grey, the New York artist who invented Zoob. It's the product of his interest in art and science, and it's become a popular learning tool as well.

I hand out the five hundred Zoob pieces to the monks, who work in groups. No, *work* is the wrong word; this is structured play. In the spectrum of teaching methods, play is the most neglected. How could anyone compare the rigor or discipline of teaching with the pleasurable anarchy of playing? Easily! Games that combine constraints with abundant possibilities are extremely useful in conveying the essence of science. The landscape is unfamiliar and liberating when there are no clear expectations and no right answer. After all, we live in a universe governed by simple laws of nature where almost anything is possible.

The room grows quiet. There's no sound except for the soothing click of plastic balls into plastic sockets, and plastic slots into plastic grooves. The monk's expressions range from soft and dreamy to furrowed concentration. They've entered a world of hypothetical physics, and they're inventing science from scratch. The abstraction is natural to them. They know the toy is a metaphor for something that can't be touched. Zoob is returning us to a deeper study of where energy comes from: atomic structure.

The Zoob pieces are like the particles in the nucleus of an atom.

Manipulating the pieces, they have a tactile way to deduce the forces that bind particles. We can also learn about the kind of rules that govern interactions of particles inside the atom. I let them play with Zoob for fifteen minutes, trying out its myriad construction possibilities. I sense they could stay engrossed for hours.

If the connections are forces, I ask, how many types of forces have you found in the way the pieces can be snapped together and pulled apart? All of the groups have found two different connection strengths. When a Zoob joint is between a ball and jaw, or a jaw and jaw, the bond is tight. It takes effort to break. This type of connection forms linear chains. But when a jaw meets the cylindrical body of a piece, or when two grooves snap into each other, the bond isn't as strong. They're both crosswise connections that allow sheets or lattices to be made. "Did anyone find another type of connection?" Some monks found an even weaker bond. If a jaw meets the body of another piece at a forty-five-degree angle, it grips but not with much confidence.

I give them another fifteen minutes to create structures using these five different Zoob pieces with their three connection strengths. Nature has four fundamental forces, and two of them operate at the scale of the atom. Zoob is a helpful tool for looking at those: the strong and weak nuclear forces. The monks link their pieces in all possible ways. Walking around the room, I see everything from tight, symmetric configurations to chaotic, rainbow tangles. It's time to be systematic in classifying our knowledge. I draw an x-y grid on the board and label the five types of Zoob pieces, as if they were particles, going down and across. Monks from each group fill in the grid and get loudly corrected by their peers if they miss something.

Here's the result: Green pieces (with jaws at either end) connect to each other only by the strongest force. Yellow pieces (with a ball at either end and a central notch) connect to each other by the medium force. All colors other than yellow have jaws and cylindrical bodies so they can connect by the weak force. And so on. All possible combinations are written on the board. We've modeled a basic principle of nuclear physics: a limited number of particles and forces determine all the structures that could possibly exist.

With Zoob we've mimicked the two primary forces that operate in the atomic nucleus. Just as I'm worrying about finding time to explain the other two forces—electromagnetism and gravity—to the monks, it suddenly happens. The monks explain it to each other. We're wrapping up for the day when Lobsang and Thupten start throwing Zoob pieces at each other. When I realize the implications, I stop everyone, and point out they've discovered photons and are modeling the electromagnetic force. In other words, photons are constantly being emitted and absorbed by atoms, and by flinging Zoob around they are modeling the electromagnetic field. That only leaves gravity. I look at the colorful chaos of a room strewn with Zoob and I'm satisfied. It's near the end of my second week. We've covered a lot of ground. The monks have fully immersed themselves in learning, and they've left their comfort zone faster than I have.

That evening, as I decompress from my day of teaching, my satisfaction is mixed with an uncomfortable feeling that I can't identify. When it comes into focus, I'm embarrassed because it's jealousy, one of the "precept violations" or sins identified in Buddhism. Why am I jealous? In an odd case of man versus machine, my twenty-question computer device, our "mystery" ball, has become more popular than my classroom lectures. The translators have had it for several days, and report beating it and losing equally often. They even rope in friends to play with it by cell phone. Apart from the million things it knows, it's programmed to be a brat: smug when it gains the upper hand, cocky in victory, and grudging in defeat. If it were a person, you wouldn't want it as a friend.

The monks are eyeing the mystery ball even more than the translators. Despite my ignoble twinge of jealousy, that evening I bring the little half-dome computer back for a special session. The first monk who volunteers selects "sound" as the item. It's a clever choice and the machine can't get it in twenty questions, so the monk wins. But the other monks have noticed that he gave conflicting answers to the questions, particularly when he said that sound could be touched. The judgment of the group is swift and unforgiving. They assail him with a loud chorus of hoots. A crashing wave of peer pressure forces him to return the NASA sticker I've just awarded him.

It's a shocking thought—a monk who cheats. Apparently, there's no end to our human struggle with precept violations.

We have time for one more round on the mystery ball. Dawa thinks for a minute. Then his face lights up. "Buddhism," he says. The machine is programmed to handle things rather than human philosophies, so it fails. Even so, its guesses along the way are surprising and a bit uncanny. After ten questions it guesses a clown, which matches the lightness I see so often in the classroom. After thirteen questions, it guesses ocean wave, which echoes the spirit of the teachings of the Dalai Lama, and its final guess before conceding defeat is friendship. Perfect.

Structure and Evolution

CENTURIES AGO scientists were called natural philosophers. Some profound scientific questions still sound very philosophical. One of these is, "Why is there something rather than nothing?" It was formally posed by the German philosopher Gottfried Leibniz, whose chief accomplishment was to invent a form of calculus. Although modern science has mostly stayed away from such philosophical conundrums, this particular question continues to boomerang back into the picture. Some scholars say we don't need an answer. Bertrand Russell, the eminent British mathematician and philosopher, put it this way: the universe "is just there."

Here in our educational enclave in India, it's clear that both Buddhism and science can't escape the something-and-nothing topic. For me, this debate is a reality check. The lives of my students are concrete and tangible—definitely something. They fled here and are in exile. At the end of that journey they encountered a modern, chaotic world, a very different place from the Tibetan highlands. We'll have plenty of time in class to talk about "nothing," cosmologically speaking. But mostly I'm compelled by the sheer somethingness of the lives of my students. Today I'm thinking of two in particular.

The first is Sonam Choephel. At home in Tibet he looked after his family's cattle. During long winter nights they huddled by the fire, saying prayers or hearing his father tell stories, many of them ancient folktales. He was illiterate until he was sixteen. Then the Chinese occupation became too threatening for his family to continue as they had done previously. Having heard of refuge in India, they began a journey across the Himalayas, aiming to pass through Nepal. At one encampment along the way they were set upon by Chinese soldiers. All Sonam remembers is soldiers shouting, dogs barking, and gunshots ringing out. Sonam and his fellow travelers ran for their lives. They ran for hours until, exhausted, they finally stopped to drink deeply at a snow-fed river.

Life was no less dangerous when they reached Nepal. While staying with a local family, three Nepali soldiers appeared at the door with rifles. So they ran again. Two dozen soldiers gave chase. They ran all night through the rain and through the rice paddies and forests. But they felt lucky. Nobody was captured or injured. Finally they faced the snowfields. It was windy and snowing as they made the crossing. They were heading into a complete whiteout. Exhausted again, they huddled together to try to stay warm and get some sleep for the night, listening all the while to avalanches booming in the distance.

Another student, Thapke Lodoe, had also made the journey. But what stands out in his memory is something else: the surprise he felt when he first set eyes on foreigners. He was eight years old. He and his schoolmates were astonished to see yellow hair and blue eyes. They were afraid to go near the visitors. Thapke thought they must be ghosts. But a friend said that was silly, because ghosts can't be seen in daylight. So they decided that the strange creatures were aliens.

Now in India, Thapke Lodoe cannot give up the sports that he enjoys so much, even though they're forbidden at the monastery. He sneaks out into a nearby forest to play soccer or cricket. He accepts the punishment when he gets caught: extra kitchen duty or sweeping the monastery's huge front courtyard. Many times he goes to the forest barefoot with a single small ball. But he has no doubts about his choice to be a monk, not a sports player somewhere else in the world. "I feel that Tibetan monks have something important to contribute to the world," he tells me. "You can't expect it from each and every monk, but there are many contemplative practices like compassion and loving-kindness that are relevant for a person living a normal life."

Compassion and loving-kindness are intangible, but I'm sure they're real and so something.

For all this something, however, Buddhism is unique in arguing that, ultimately, there's nothing. Everything is an illusion of human invention. Behind their lives of something, the monks accept that there is nothing. For all their harrowing adventures and personal pleasures, reality finally is an empty void. It is an idea that has been chilling to Western religion, and even to the Western existential psyche (that's why Leibniz, among many others, asked the question, and then offered his metaphysical assertions about God's existence). That was not the solution for Buddhism. The Void, not God, is its ultimate answer. As we're about to learn, even science hasn't escaped this philosophical conundrum.

Something Rather Than Nothing

Christianity and Buddhism have always understood the terms *nothingness*, *void*, and *nonexistence* differently. Early Christian theologians didn't accept the Greek and Near Eastern mythologies that saw the world arising from diverse gods or from a primeval form of matter called *chaos*. Drawing on the Hebrew idea of a Creator who brings chaos under control, Christian theologians went even further with the idea that God created *ex nihilo* (out of nothing). In this sense, everything that exists depends on God at every moment, and by the same token, *nothing* is the absence of God and to be abhorred. Buddhism, by contrast, assigns no omnipotence to a deity. Instead of heaven, there's nirvana, the "blown-out candle." Instead of eternal personal afterlife, there's extinction of the self. The Void is embraced.

Modern cosmologists have been mindful of the Christian and Buddhist definitions, and also the fanciful creation stories of ancient cultures. One of my students, a monk named Sherab Tenzin, said he once met a Western physics teacher who believed God created the universe, and he asked Sherab what Tibetan Buddhists believe. "I teased him," Sherab said. "I told him people are made in the shape of a human and fried in a pan. The first group of people was all burnt and came out black, so the god sent them to Africa to become the African people. The next time he took people from the oil they weren't cooked very well, so they became the white people. Now the god had learned how long to keep them in the oil, so finally they came out perfect and cooked properly. They were the Asians."

The wide variety of creation stories, serious or ludicrous, has made it even easier for scientists to stay away, or at least limit their own concerns to Spartan versions of the Leibniz question. What was the initial state of the big bang? How did matter form? Why is the universe expanding? The scientific answers have been many. But one of the most compelling in modern times has come from British cosmologist Stephen Hawking. It's called quantum creation. He has elegantly summarized the idea in one of his best-selling books, *The Universe in a Nutshell.*

The theory goes like this: In the very early universe, the big bang started due to a quantum fluctuation, the smallest of jolts, so to speak, at a subatomic level. Though small, it triggered the creation of energy and matter. The universe expanded suddenly and rapidly by borrowing energy from the vacuum of space, essentially from the void. From this energy, matter arose. This initial rapid burst has been called "inflation," since the universe abruptly increased in size by many orders of magnitude. Oddly, the total energy is still zero because the positive matter energy is exactly balanced by the negative gravitational energy. The total stays at zero even as the universe grows in size. In other words, it has become a very substantial something out of virtually nothing,

After inflation, the universe expands at a more sedate rate. It eventually takes on the form we see today—large and nearly empty, with flat space-time. Nothing has been added or removed, so the total energy is still zero. Positive expansion energy is balanced by the negative potential energy of gravity. I'm waiting for the inevitable question, for even the Buddhist mind, like the mind of Leibniz, wants to know about causes. Tashi Phuntsok puts his hand up. He asks, "What caused the quantum fluctuation that caused the universe?"

I smile. I'm tempted to say that scientists don't go near such philosophical questions. What I do say is that the scientific story of the big bang doesn't answer the question of genesis. It points back to a state of affairs where current scientific theories lose their traction. This corresponds to a time when 100 billion galaxies were squeezed into a region smaller than a proton. This iota of space-time is called the Planck scale, in a nod to one of the founders of quantum theory, the German scientist Max Planck.

This limit on our physical measurement of the early universe is characterized by fantastic numbers. Imagine the universe when it had a density of 10^{90} kilograms per cubic centimeter and a size of 10^{-35} meters, a mere 10^{-43} seconds after the big bang. Science reaches its limit here because the four forces of the universe were likely melted together into a single superforce. There's no good theory of that superforce because we don't know how to reconcile quantum theory and gravity theory. I stress to the monks that the scientific method is on thin ice when it addresses the origin of everything.

The challenge hasn't stopped theoretical physicists from reaching deep into their tool kit and trying to say what caused the big bang. The key presumption is that the four forces we see in the universe today were unified at the very beginning. Unfortunately, the first two forces—the strong and weak nuclear forces—and the second two—electromagnetism and gravity—couldn't be more different from each other. The first two don't extend beyond the atom. The second two have infinite range. Also, in our current large and cold universe, these forces differ in strength from each other by a factor of 10^{38}.

A breakthrough in our understanding came in the 1970s. Physicists developed a theory to relate electromagnetism and the weak nuclear interaction. Then hints of the underlying unity were seen using particle accelerators. The new theory predicted that at 1,000 trillion degrees, a paint-blistering 10^{15} Kelvin, the two would melt into a single force. That temperature occurred ten trillionths of a second, 10^{-11} seconds, after the big bang. The experimental verification of this unification led to the award of the 1979 Nobel Prize in Physics.

Emboldened by this success, physicists reached for the next level of unification, where the strong nuclear force is added to the mix. This is the "grand unification" challenge. The physics is challenging because the particles that interact by the electromagnetic and weak nuclear forces have quantum spin numbers that are integers, while the quarks that interact by the strong nuclear force have half-integer spins. They're immiscible fluids, like oil and water. In grand unified theories the three forces only melt together at the fantastic temperature of 10^{27} Kelvin, which occurs a minuscule 10^{-35} seconds after the big bang. Such temperatures, I tell the monks, are so outlandish we might as well be speaking of magic.

Suddenly we segue from the topic of temperatures that melt the universe together to how we can talk about "seconds" or time in that realm. This new line of inquiry begins when my stalwart interlocutor, Thupten B, raises his hand. He passes his palm over his shaved head as he asks his question. His face is earnest and thoughtful. Tenzin translates. "If the most fundamental way to measure time uses the oscillations of atoms, how can we measure time when there are no atoms?"

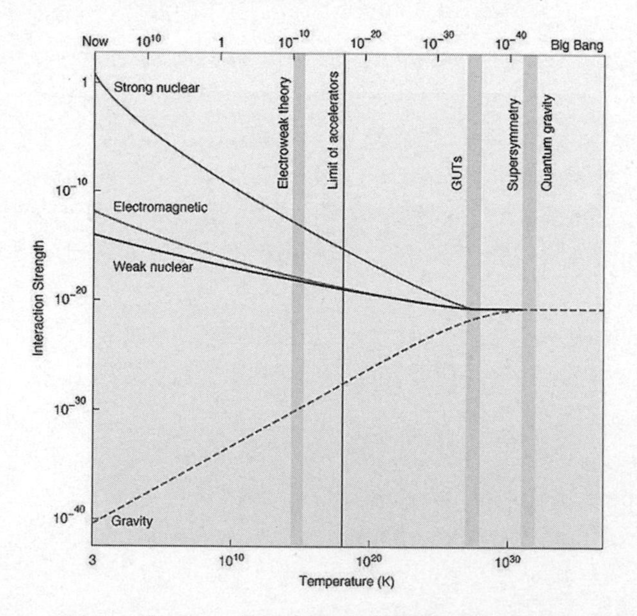

Figure 5.1. The unification of the forces of nature is presumed to be realized at a temperature far beyond the reach of human technology, but this state occurred very early in the history of the universe.

A very good question. "We can't," I admit. Time is purely a hypothetical construct in the early universe, a creature of theory. In other words, time at these very early epochs depends on extrapolating general relativity into regimes where it's never been tested. Assuming general relativity is correct, every physics classroom on the planet, including ours in India, can take a deep breath, be awed by the enigma of no time at the beginning, but nevertheless continue to try to build the scientific theory of creation. Grand unification physics is still in its infancy.

Meanwhile, I assure the monks, "There's no single theory. Dozens of variants have been proposed. But most theories share one common feature. They assume symmetry, which is probably the most fundamental premise of physics. Symmetry is expressed mathematically but like the best pure math, it has an aesthetic quality. Symmetry in this case is a kind of equivalence of all particles and all forces at the beginning. The four forces were equal—there were equal numbers of particles and antiparticles—and all members of the particle "zoo" behaved the same way. However, we don't see symmetry in the diverse world of subatomic particles now. So physicists have had to explain the source of the nonsymmetry. They've proposed "shadow" particles to make

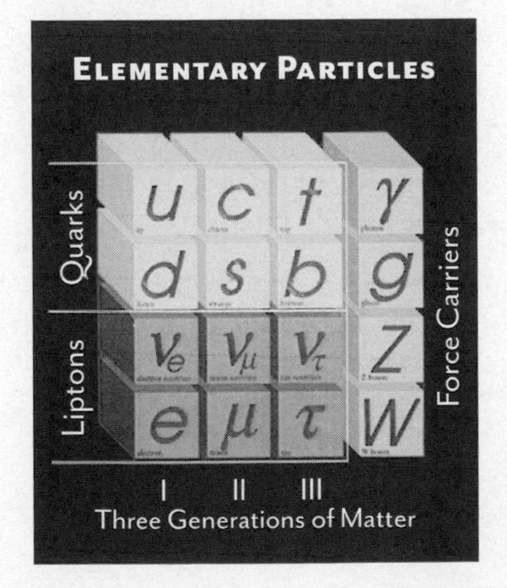

This scheme has multiple generations of particles and their antiparticles, so it is not very elegant or simple. This has led physicists to suppose that there may be an even deeper level of subatomic structure.

Figure 5.2. The arrangement of particles and forces in the "Standard Model" of particle physics is unsatisfactory because the details of the scheme aren't specified by the model, implying a deeper level of physical reality.

up for the imbalance. It's inelegant to postulate extra particles when there are too many to explain already. But that's the cost of the elegant organizing principle of symmetry.

To help the monks imagine this idea of symmetry when, in homely language, everything was "melted" back together, I need an analogy. The spinning coin example will do just fine. I point to our classroom floor and say, "Imagine a set of coins all spinning on the floor." This would be too difficult to demonstrate perfectly, so we use our imaginations. When the coins all spin, they are in a high-energy state. They have symmetry—or equality—by all being on their edges, neither heads nor tails. When they lose energy they fall into one of two distinct categories: heads or tails.

Theoretically, if we could generate enough energy, we could get all the particles and interactions to go back to perfect symmetry. We could melt everything back to a unity. In the realm of unity, all enigmas are solved: we'll understand how to reconcile general relativity and quantum theory and, more broadly, how the four forces of nature began as a single *superforce*. In reality, re-creating that moment of unity is impossible. The limit of current particle accelerators is a temperature of 10^{17} Kelvin. This is 10 billion times cooler than the energy needed to produce the original symmetry. The largest accelerator in the world, the Large Hadron Collider in Geneva, can only glimpse that holy grail of symmetry and unification.

Yet in physics, hope springs eternal. Even though we don't have the power to re-create the original symmetry, we have bits of evidence, logic, and mathematics to imagine how it worked. This is the scientific quest for a "theory of everything." At this level, physicists treat us to a wild garden of ways to explain how it was at the very start. Some may sound exotic to nonphysicists. As we've seen, one popular idea is that particles are actually one-dimensional "strings." These strings oscillate and vibrate to produce the effects that we see as particle interactions. Successful versions of the theory involve invisible extra dimensions of space.

A related idea is that the vacuum at the start of the universe could have had a vast number of potential energy states. That gives us a basis in physics to suppose that the patch of space-time that inflated to become our universe was just one among many. An infinite number of universes may have emerged from the vacuum. Each one operates under different laws of physics. These universes are distinct space-times, unobservable by us. This is the "multiverse" hypothesis.

"But is it true?" asks Thupten. It's a simple and disarming question. I want to come to the defense of all theories of everything, because they're popular among very smart physicists, and they're deemed to be plausible speculation. They're also motivated by the fact that the standard model of particle physics is incomplete. Something like string theory fills the gap, as does the multiverse, which answers other questions, such as, why did the cosmic roll of the dice give our world its particular laws of physics? But I hold back on my full-throated defense. I know how attached we scientists can be to our theories, which serve as springboards to careers and even fame.

Here in India I'm attached to explaining difficult things. The best scientists know how smart they are and that's an attachment, too. This non-Buddhist strain in science gets in the way of progress at times. A herd mentality means that people work on some problems and ignore others. It's mutually affirming if everyone agrees that string theory and the multiverse are "where it's at." There's been a backlash against both ideas by some well-respected scientists and philosophers. Without testable predictions, these clever ideas can't enter the mainstream of science. "We don't know," I honestly answer Thupten. "We just don't know."

At their base level, scientific explanations run into metaphysics or philosophy. "Why are there laws of physics?" I tell the monks that we don't know why these laws are the way they are. Postulating alternative universes with different laws of physics doesn't explain our laws of physics, and it doesn't explain *why* there are laws of physics, just as quantum genesis doesn't explain *why* the universe came into being. We haven't answered the original question: Why is there something rather than nothing?

I assure the monks that science and Buddhism share an uncanny similarity in approaching the natural world. Both state that the physical world is rational and governed by cause-and-effect. As a philosophy, Buddhism further posits the precept of dependent origination. All things are dependent on something else. This resonates well with quantum physics, but it's also consistent with frontier theories in cosmology. In Buddhism, "universes" are continually disappearing and coming into being, which was perhaps the situation at the Planck era.

Nonetheless, our well-intentioned quest to identify an ultimate cause may be doomed to failure. It does verge on absurdity, as illustrated by a story the astrophysicist Stephen Hawking told in his blockbuster book, *A Brief History of Time*. A famous scientist gives a lecture and talks about how the Earth goes round the Sun, and the Sun is part of the Milky Way, and the Milky Way is one of many galaxies in the universe. A lady at the back of the room stands and says, "What you've told us is rubbish. The world is really a flat plate supported on the back of a giant turtle." She's alluding to an ancient origin story, perhaps of Hindu extraction. The scientist gives her a superior smile and asks, "What is the turtle standing on?" "You're a very clever young man," replies the lady, "but it's turtles all the way down!"

Monk Gravity

Turtles notwithstanding, we've been pretty esoteric in class today. It's time to get more substantial, which means some role-playing among the monks. We've used marbles, flashlights, and fluorescent cutouts before to be galaxies or photons, but now it is time for a full-dress performance. *Lucha libre* is a form of masked wrestling popular in Mexico and other Latin American countries. This burly wrestling tradition will help us review the four fundamental forces in the universe.

As it happens, I have a set of lucha libre masks with me. I ask for four volunteers to act out the big bang, an event in which the four fundamental forces reveal themselves. Dawa, Jigme, Thupten, and a burly monk named Ngawang Sherab step forward. As I hand them four colorful masks I give a little background on this exotic Mexican import. Lucha libre means "free fight." The most successful wrestlers are folk heroes, where they travel the countryside and compete in rings set up in dusty town squares. The mask of a lucha libre wrestler becomes his identity, and it can be styled on animals, gods, ancient heroes, or other archetypes. When a wrestler is defeated and unmasked, he loses that identity and must find another, or he must wrestle without a mask.

We're going to morph this folk tradition into laboratory physics. The analogy is not too far off because the beginning of the universe was like a gigantic wrestling match between energy and matter.

The monk who will present the strong nuclear force is Dawa, and he wears a red and blue lucha libre mask. Ngawang gets green with purple trim, and he's the weak nuclear force. As electromagnetism, Jigme dons a purple mask with yellow lightning. Thupten gets the last mask, colored red and black. He's playing gravity. The masks were made for children, so they're a tight fit, especially for big-boned Ngawang and Thupten. Jigme flexes his rather modest muscles and the others laugh.

I worry that I'm imposing frivolity on these serious monastics. I needn't have worried; the monks are nothing if not good sports. For a moment they're self-conscious, but soon they're mugging with the new head gear. The demonstration is elaborate, so we block it out as if we were plotting a scene in a play. My actors master their parts, and I add the final props to the scene: a pile of Zoob on the floor and a flashlight for Jigme. We're ready. The other monks take their seats on the floor with a rustle of anticipation. This is what they see.

Our four masked actors stand in a group facing inward. They tilt their heads down and lock arms around shoulders, like a small rugby scrum. I clap loudly. *Bang!* The universe springs into action. The clump of wrestlers starts moving in a circle. Round they swirl. Faster and faster, one seething mass of burgundy cloth and brown flesh, until the superforce holding them together breaks and Thupten is flung off and tumbles to the floor. The remaining three, with locked arms, continue to swirl. Thupten, playing gravity, crawls on his belly as if pinned by an implacable force. As gravity weakens in the first iota of time, Thupten is able to get on hands and knees. Then he's up, staggering like a drunk.

Meanwhile, the scrum ejects a second wrestler. This is Dawa with the bold red and blue mask. As the strong nuclear force, Dawa approaches the big heap of Zoob on the floor and starts making tight little constructions—the nuclei of atoms. Jigme and Ngawang continue to spin until they break apart, flinging Ngawang into the front row of monks. Ngawang is dizzy, but he lurches over to Dawa and starts breaking up his Zoob nuclei. Dawa fends him off. They tussle like genuine wrestlers, with Ngawang trying to undo the Zoob that Dawa is putting together, and then putting him in a headlock when Dawa ignores him. Playing the role of the weak nuclear force that causes radioactivity, Ngawang knows that his job is to break atomic nuclei.

Meanwhile, Jigme is the thespian playing electromagnetism. He wanders around the room shining his flashlight in people's faces. I glance at my watch. It's a trillionth of a second after the big bang.

"A tale told by an idiot, full of sound and fury, signifying nothing." So said Shakespeare's Macbeth. In our theatrical presentation, the universe has seen a lot of action, but the "something" that it's made of hasn't turned into anything. In the first fraction of a second it's a cauldron of radiation and particles and antiparticles coming in and out of existence. Then at some point it's no longer hot enough to create matter and antimatter spontaneously. From here the very slightly larger number of particles annihilate the very slightly smaller number of antiparticles, which creates a flood of radiation. It also leaves a residue of particles, which will be sufficient to eventually forge 10,000 billion billion stars.

In the last act of our wrestling drama, Thupten returns. Having been ignominiously flung aside in the first act, he has bided his time as he gathers in strength in the growing universe. About ten thousand years after the big bang he seizes his chance. For the first time, radiation no longer dominates, and the weakest of the four fundamental forces starts to exert its influence. Gravity has arrived. This will be the height of Thupten's career in stage, screen, and lucha libre. He adopts a bodybuilder's pose, fists clenched and arms flexed. Wearing his red and black mask, he began as the weakest force among the four, but now he will shape the universe.

Gravity has become our topic, and it will require more technical and historical background than even Mexican wrestling can offer. For the story of gravity, we get acquainted with two titans of physics: Isaac Newton and Albert Einstein.

Newton was an enigma, a public scientist yet a personal cipher, with a formidable intellect. Listen to Richard Westphal, who took twenty years to write the definitive biography of Newton: "The more I have studied him, the more Newton has receded from me. . . . He has become for me wholly other, one of the tiny handful of supreme geniuses who have shaped the categories of the human intellect, a man not finally reducible to the criteria by which we comprehend our fellow beings." Now that's praise.

Newton admitted that he stood on the "shoulders of giants," but he single-handedly came up with a theory that explained all astronomical orbits of his time, described the behavior of as-yet-undiscovered stars and galaxies, and still provides an excellent tool for studying almost all motions in the universe. Our lucha libre wrestling is over and I'm at the whiteboard, writing out Newton's universal law of gravity. It's simple: the attractive force between any two objects is proportional to the mass of each one and inversely proportional to the square of the distance between them.

NEWTON

Mass and energy are independent and are very different things.

Space and time are independent and are very different things.

EINSTEIN

Mass and energy are linked and they are interchangeable: $E=mc^2$.

Space and time are linked and are part of a four-dimensional space-time.

Figure 5.3. Newton and Einstein used conceptual frameworks that were completely distinct. Newton imagined that space and time were linear and absolute; Einstein saw them as supple and interrelated.

I note the important features of the equation. The equality means that the force of A on B is the same as the force of B on A. We tug upward on the Earth with exactly the same force as the Earth tugs down on us. That reciprocity is worth remembering on days when it's hard to get out of bed. Enough people in one place jumping at the same time really would move the Earth by a tiny amount. The force has infinite range because it declines as the inverse square of the distance—one over the square of a very large number is very small but never zero.

As we've discussed before, Newton's theory is commonly misunderstood to imply a "clockwork universe," where the motion and position of particles can be predicted with perfect precision. Humanists of the Romantic era railed against the theory, because determinism seemed an assault on free will. But the situation of two objects is artificial; in the real universe situations are always complex and involve the influence of more than two objects. At some level everything attracts everything else—a very Buddhist concept.

Newton lived before anyone had measured the distance to the stars. The discovery of galaxies and the expanding universe were far in the future. He thought the universe was infinite and eternal. When he was asked, "What's this force that acts instantly and with infinite reach across the vacuum of space?" he essentially shrugged and said, "I have no idea." Newton wasn't always taciturn. For instance, he was inclined to say that when orbits of planets got out of sync, the Creator simply adjusted them. This led to a great debate on the role of God in the clockwork universe between Newton and Leibniz, who were already at loggerheads over who was first to invent calculus.

Newton's accomplishment in gravitational theory presented another problem, involving gravitational collapse. To demonstrate, I call on the monks to create a world of gravity in action. I get everyone to stand up and push the cushions to the edge of the room. I ask them to stand about arm's-length apart, randomly and not in rows or any regular pattern. When gravity switches on, I tell them, "Reach out and pull the nearest monks to you closer." The monks will be acting as particles. Our two-dimensional floor will stand for three-dimensional space. And their pulling will represent a universal attractive force that extends in all directions.

Trying not to let the power go to my head, I activate the experiment. "Gravity go!" I hear sandals shuffling on the wooden floor and see plenty of tugging. From the front I can't get a view of everyone. "Gravity stop!" I ask a couple of the monks near the center how much the pulling has moved them. "Not far," says one. "Nowhere," says another. Geshe Nyima is at the edge of the group on my side. Where did you move? He points toward the center of the group. We switch on the force again: "Gravity go!" There's more shuffling and tugging. "Stop!" By now it's clear that the entire group is closer together. After one more burst of gravity the entire cohort occupies half the space it did when we started. Voila! We've simulated gravitational shrinkage.

Breathless and sweaty from play-acting gravity, the monks and I pause and regroup. Why did the monk "cloud" collapse, we ask ourselves, and what does that imply for the universe? Geshe Nyima offers the best analysis: monks at the edge of the group are only pulled by monks closer to the center of the cloud. So they move inward. Monks near the center are pulled by neighbors on all sides so they generally don't move. The cloud shrinks from the outside in, and as the distance between monk "particles" gets smaller, the gravity force increases, so the shrinking continues.

In fact, I explain, the shrinking gets faster. It becomes a runaway phenomenon called *gravitational collapse*. I pose a question, "What if the monk cloud is infinite?" This causes lots of discussion. The monks go into their monastery groups to discuss this thought experiment. After a while, several groups offer an answer. With an infinite sea of monks, each one is pulled in all directions equally. They may move around a bit but they won't go anywhere. Since no particular monk moves, the whole collection doesn't move. An infinite universe must be stable and static.

"Exactly!" I'm impressed with their logic. Newton knew that a finite universe would be unstable, so he preferred an infinite one. Even so, Newton's infinite universe raises other logical problems, which have to do with the behaviors of gravity and light. Each declines according to the square of the distance from the source. But moving out in space, the number of sources of light or gravity in successively distant regions increases as the square of the distance. So the contribution of light or gravity moving outward is constant. In an infinite universe, the amounts of light and gravity should be infinite. That's a real problem! It would take the next titan of physics, Albert Einstein, to resolve the anomalies. To explain how Einstein took the baton from Newton, we pick up from where we left off in our story of cosmic evolution.

The Way of Newton:

Mass tells gravity how much force to exert,
and that force tells the mass how to move.

The Way of Einstein:

Mass-energy tells space-time how to curve and
curved space-time tells mass-energy how to move:
theory of general relativity.

Figure 5.4. Newton's universe was infinite, which posed problems for the total amount of gravity and light. Einstein's theory incorporated space curvature, and a curved universe can be finite yet unbounded.

It's ten thousand years after the big bang. Gravity (aka Thupten) is beginning to flex its muscles. The next landmark is after 380,000 years, when the universe was a thousand times smaller and hotter than it is today. Stable atoms form. Radiation starts to travel freely. After all the backstage pyrotechnics, this is the curtain raising of the universe. Anyone present at the scene would see space filled with dull red light. I tell the monks, the universe once glowed the same color as your robes! We've talked about this era before; the radiation that was later redshifted to microwaves is the strongest evidence that the big bang actually happened.

The ripples in the radiation from the big bang are tiny, just one part in a hundred thousand. That's like having a pond a hundred meters across where the surface is ruffled by a light wind and has ripples a centimeter high. "But big oak trees from little acorns grow," I say, waxing poetic (and hoping Tenzin can translate my mixed metaphors). Over the next few hundred million years, as space silently expands and the universe cools, the little ripples grow into waves. The lights go on in the universe as the first galaxies form. Each one forms similarly to the collapsing monk cloud experiment we did earlier. The galaxies when they form are randomly distributed in space. But I've shown pictures of the way galaxies in the present-day universe are distributed. They congregate into groups and clusters, with great voids between them. How did this happen?

We've tried gravity wrestling, but compared to that, what we're about to attempt is quite audacious. We're about to launch the Monk Gravity Project. I believe my monks can do it because of their remarkable patience and concentration. At home I would be making a totally unreasonable request of my students, but what we're about to do seems custom-made for monks. I sense that their sheer curiosity about gravity will carry them through.

For this activity we need a large room. Bryce tells me about an unused auditorium on the ground floor of our building. With one look I realize it's perfect. I ask the monks to come for an after-dinner session. We all arrive, and I'm lugging a bulky bag filled with ten thousand black plastic poker chips. I have a similar number of small, circular blue stickers. The Monk Gravity Project will take a few sessions, so on this first evening, we're going to learn the basics of the inverse square law of gravity. Over the next few evenings, "We're going to simulate the evolution of a sizeable chunk of the universe," I tell them. If we do well, we'll see with our eyes how gravity has crafted the large-scale structure we see in galaxy surveys.

The inverse square law, which I had written on the whiteboard, was Newton's idealized formula. The monks understand, however, that its simplicity had to be expanded. Newton spoke of just two objects, but in the universe many objects are pulling on each other at the same time. Back at the whiteboard, I show how this works with three or more objects. I draw a series of arrangements of three, four, five, and six circles, where the circles are supposed to represent galaxies. Later, in our hands-on experiment, poker chips will serve as our galaxies. On the board, I highlight the circle nearest the center of the arrangement. Then I ask, which way will this galaxy move in response to the gravity of the others? Everyone offers opinions. The actual answer is ambiguous, because the movement will be a dynamic trade-off between the distance to the neighbors and their relative orientations.

Generally, the closest neighbor dominates the local gravity, but two or three neighbors slightly farther away and in a different direction can balance or exceed the force of the nearest neighbor. It's what mathematicians call a "vector addition" problem. The monks and I practice this type of addition with a few easy cases. Then we work up in complexity. Each case is a miniature gravity puzzle. We're doing simplified chess puzzles in preparation for the game itself. After two hours, I'm confident they've got the inverse square law lodged in their intuitions, and we conclude a very long day.

We're back in the auditorium the next night. Anticipation is in the air. We are about to simulate the origin of the universe and how gravity formed the galaxies. The ten thousand poker chips are out, and we've decided to add some pageantry. With bags of chips in their hands, the geshes are the benevolent creators of our universe. They walk across the auditorium stage, tossing handfuls of poker chips high into the air, like farmers sowing seed. We all watch as the plastic disks skitter and roll to a stop. I tell everyone that the starting situation should be as close to random as possible, so we make one pass over the floor, stirring the chips with our hands until the floor is a featureless sea of black dots. Next I hand out the blue stickers, a hundred per sheet. Yes, it's true, I ask the monks to put a blue sticker on each of the ten thousand black chips, which are strewn everywhere. This will allow us to keep track of our experiment since we can turn chips over, creating a blue or black code.

When this is done, we ponder the fact that at our feet is a two-dimensional "model" of the universe. It's a half a billion years after the

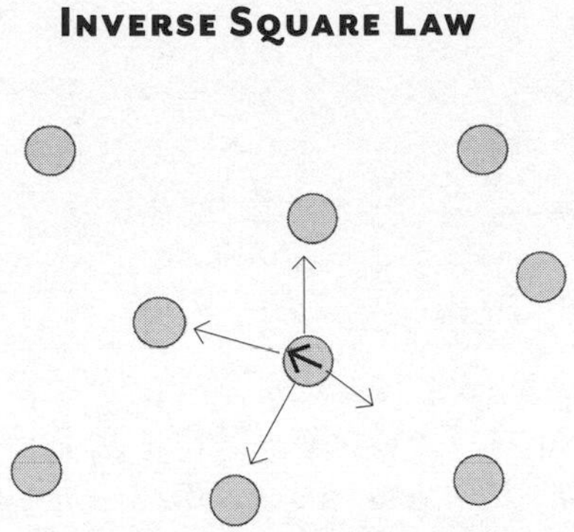

Inverse Square Law

Figure 5.5. To prepare for simulating the large-scale structure of the universe, the monks study examples of two-dimensional gravity. Each galaxy moves in a particular direction (bold arrow on the disk) because of the pull from neighboring galaxies (arrows pointing at nearby disks).

big bang. The galaxies are young. A region with ten thousand galaxies is about 300 million light-years on a side, and it's been shrunk into this dingy auditorium in Himachal Pradesh. The monks are controlling a significant chunk of the universe. Researchers do what we're doing with computers. They set up a virtual space with smoothly distributed matter and calculate the force of every galaxy on every other galaxy. They move them all according to Newton's law; then, as the positions of galaxies change, researchers repeat the process. With the lightning speed of billions of calculations per second, the computer simulates the evolution of structure and the gradual emergence of clusters, filaments, and voids. In this remote part of India, with few computers and flaky electricity, we'll do it the hard way.

We begin. The monks move across the floor methodically. Barefoot, crouched on their haunches, they squat over each chip and scrutinize the configuration of nearby chips. Then they move each chip slightly according to the sum of all the forces acting on it. Once a chip is moved, it's turned over so the blue dot is hidden. The monks work with delicacy and dedication, and in this first round, which takes an hour, no chip is left unturned. The sea of blue dots has turned into black. We retreat to the perimeter. To the eye, very little has changed across the auditorium floor. The chips are now all black, but the pattern still looks random.

"Okay, again!" The monks float across the chips a second time, adjusting their gravitational relationships. As some chips are moved closer to each other, they are moved away from others. Each pass requires thousands of judgments about gravity. Still, after two passes, there's no dramatic difference in how the scattered chips look on the auditorium floor. I wouldn't dream of asking my students back home to do this experiment. But the monks approach it cleansed of attitude or expectation. I think of the piles of black pebbles I saw at the temple in Dharamsala. Buddhist pilgrims use little stones or seeds to count the number of times they've walked clockwise around the temple.

After the second pass we stop for the night. The third evening we're back again. With ten thousand chips to consider, the monks work in parallel for the most part, like a modern computer. At times they collaborate. Groups form to consider particularly tricky configurations of galaxies. During the third pass, a spirited debate breaks out. All the monks congregate around Jigme and Sonam Choephel, who are disagreeing about how best to move one particular galaxy. I let them debate and argue for ten minutes, then realize we'll never get the job done with so much attention paid to one galaxy. "Stop!" I say. "You're taking this too seriously." I stare at them sternly. Karma translates, they grin, and we're back at hunching over the black chips.

By the end of the fourth evening we've made seven passes over the chips. The monks have simulated several billion years of cosmic evolution. Now we can see a dramatic difference. The chips have concentrated into groups and piles. For the first time, there are clear patches on the wooden floor. What was once random has crystallized into clusters and voids. Monk gravity has built the architecture of the cosmos. It's taken twenty hours and thirty-five people to do for ten thousand galaxies what a modern computer could do for a billion in a few minutes.

No matter. We've absorbed gravity deeply into our pores, and we'll never be the same again.

At the end of the last session, I'm slow in gathering my teaching materials. All the monks and translators have left. I go to the door and to my surprise the ornate brass door is locked. I'm trapped. There's nobody around; they've all gone to bed. In exasperation I yell and

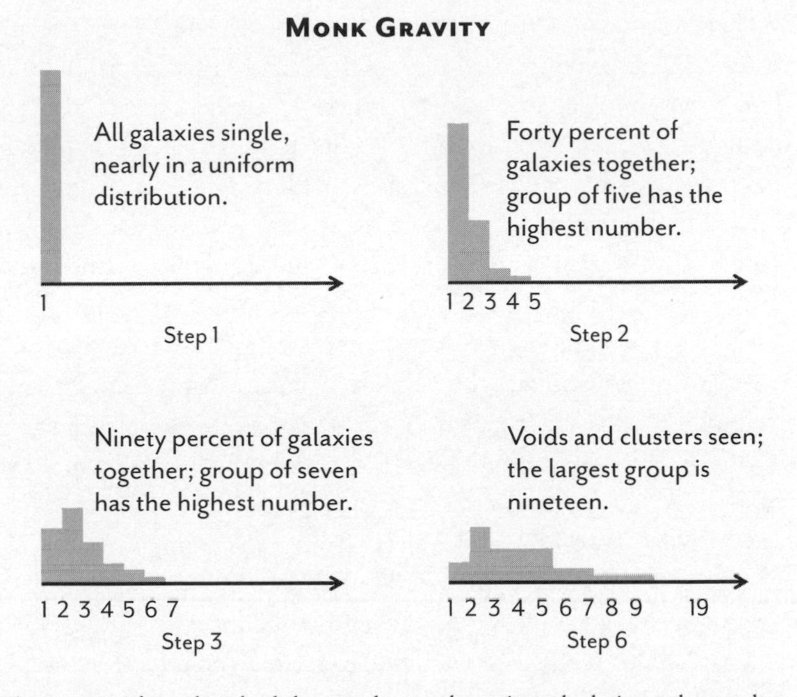

Figure 5.6. After a hundred thousand mental gravity calculations, the monks "evolve" clusters and voids. The histograms show the number of galaxies that are isolated (one in a region) or clustered (two or more in a region).

pound on the door. It looks like I'll be sleeping on a hard wooden floor. I pace the room in frustration, and finally through a window I notice Lhakdor, Karma, and Paldon off to the side. They're pressed against the wall, hiding, like mischievous children. The venerable Lhakdor is behind this, I'm sure. Bryce tells me later, they're showing you love.

Einstein Pays a Visit

Having learned about Newton by his equation on the whiteboard, and by its application to ten thousand poker chips, we turn to Einstein. The founder of relativity and solver of the gravity enigma is not to be taken lightly, I tell the monks. They will learn later that Einstein was actually a practical joker. But we start out in seriousness and awe toward the great man. I tell them, "Like most scientists, Einstein disliked coincidences."

That bears explaining. When seemingly unrelated events or processes coincide or return the same result, scientists get suspicious. They think a law must be at work. Of course, coincidences can also arise out of pure chance. I showed this on our first day in class. We found that three pairs of monks shared the same birthday, even though with thirty-four monks and 365 days in the year, that would seem unlikely. However, we were not discovering some hidden law of birthdays; the result was consistent with a random distribution of birthdays.

The story was different for a predecessor of Newton, the Italian scientist Galileo Galilei, who became suspicious of a coincidence he noticed. This was the fact that objects fell at the same rate of acceleration, whatever their weight—be they a cannonball, a horseshoe, or a small stone. Galileo suspected that a law of nature was at work. As an aside, Galileo never dropped objects from the leaning Tower of Pisa. It's a great story, made up by one of his students (just like the story of Newton seeing an apple fall and inventing the theory of gravity).

What Galileo really did was roll objects down an incline in his laboratory. By measuring their motion he showed that no matter how heavy or light a rolling object, it accelerated at the same rate. What seemed a coincidence turned into Galileo's law of motion: motion caused by gravity is independent of the amount and type of stuff in the object. Galileo knew his discovery was important because there are two conceptually different types of mass. One is inertial mass, the resistance of any object to a change in its motion caused by any force, given later by Newton's second law of motion. The other is gravitational mass, the mass that describes the gravity force between objects, described later by Newton's famous gravity law. There's no obvious reason the two masses should be equal. But by Galileo's measurement they were the same to within 1 percent.

More recently, *Apollo* 15 astronaut Dave Scott did Galileo's experiment without the confusing effect of air resistance. Millions of television viewers saw a hammer and a feather hit the lunar soil at the same time. More recent experiments have shown that gravitational and inertial mass are equal to within one part in a trillion.

So we arrive at Einstein. He had the benefit of the laws discovered by Galileo and Newton, both of whom left behind puzzles and anomalies. Einstein was well suited to probe into the most difficult areas of early twentieth-century physics, He was a scientific outsider. As a five year old, he had been mesmerized by the invisible forces that turned a compass needle. Yet by age sixteen, he was a high school dropout, draft dodger, and unemployed. All his applications to enter a university or teacher's college were rejected. So he considered himself lucky when he got a job as a technical officer, third class, in the patent office in Berne, Switzerland.

With lots of free time to think, he came up with mass-energy equivalence, the formalism of light as a photon, and the special theory of relativity, all published while he was only twenty-six. Two years later, he was sitting in his office when he had an insight he later called "the happiest moment of my life." His thought: if he was freely falling in a gravitational field, he'd be unable to feel his own weight.

Why would Einstein be so happy as he contemplated plunging to his death? Because his insight explains Galileo's observation that inertial and gravitational mass are equal. Earlier, Einstein's special theory of relativity had dealt with the case of uniform motion. In his 1916 masterwork, he was able to "generalize" the theory to account for nonuniform or accelerated motion. As a result, he removed the distinction between a change of motion caused by gravity or by any other force.

I realize this isn't easy material for the monks. So I reach for the thought experiments that are often used to explain the theory, where someone in a sealed, windowless environment has to figure out their situation.

I ask the monks to consider two situations. In one, they're inside a space capsule with no windows in deep space, far from the Earth or any other planet. With no local gravity, they would be weightless, floating freely inside the capsule. In the second situation, the same capsule is plunging toward the Earth with no parachute. The second situation sounds very different and much more dangerous, but Einstein's exciting insight was that there's no way anyone could distinguish the two situations from inside the capsule.

Now I walk around the edge of the room and close all the curtains.

Rubbing my hands with glee I say, "Evil aliens have quietly hitched this room to their spaceship. They're heading through deep space to their planet to conduct medical experiments on monks." Cleverly, their spaceship is accelerating at 9.8 meters per second per second, which is the same rate at which something you drop would accelerate toward the ground. That way, the aliens think, you won't be suspicious that anything's wrong. Fortunately, the aliens will let you go if you can prove that you are being dragged to your death rather than sitting in a classroom not going anywhere.

Jigme thinks of a solution and heads for the nearest window. No! I block his path. You can do anything except look out of a window. The monks discuss the aliens' demands in groups. The puzzle stumped Einstein, so I'm fairly sure it will stump them. Indeed, there's no way to distinguish acceleration due to gravity from acceleration due to any other force. This awareness propelled Einstein to his crowning achievement, a gravity theory that received its initial confirmation three years after he published it in 1916. It has been affirmed repeatedly in the century since.

General relativity is a geometric theory. There's some very difficult mathematics at its core—tensor calculus—which can strike fear in the hearts of mere mortals (and it even gives me the wobbles). Fortunately, general relativity can also be summarized in simpler language by contrasting it to Newtonian gravity. In the Newtonian world, space and time are distinct, linear, and absolute. In Einstein's theory, space and time are linked and supple. The equations of general relativity relate the curvature of space-time to the local mass-energy density. As the physicist John Wheeler put it, "Mass tells space-time how to curve, and space-time tells mass how to move."

Space isn't empty, and it's not flat either. The most common analogy is a rubber sheet that bends when a mass is present. Objects rolling across the rubber sheet are deflected by the curved space. A black hole is a region of space with such strong curvature that it's invisible because it's "pinched off" from the rest of the universe. Visualizing an image of the curvature in three dimensions is harder.

There are other bizarre consequences to general relativity. Mass and energy are equivalent, so photons should respond to space curvature as well as particles and planets. Let's return to the space capsule. If it's floating in deep space, a light beam shone directly across it travels without deflection. But if the space capsule is being accelerated, the light beam follows a curved path, since the capsule moves while the light is traveling.

The logic of relativity means that the same thing should happen when the capsule is stationary on the Earth's surface. The light beam bends toward the Earth because the Earth is bending space. It's a tiny effect but it has been measured. Light loses energy as it struggles against gravity, an effect that is observed as a redshift of the radiation leaving a massive object. Since space and time are linked, the distortions of space also affect time. Clocks run slightly slower when gravity is stronger. In the 1970s, physicists took atomic clocks up in U2 spy planes while identical clocks stayed on the ground. The Earth acts as if all its mass is concentrated in the center, so gravity gets slightly weaker as you move above the surface. When reunited, the high-flying clocks had gained by tens of nanoseconds compared to their Earth-bound siblings.

Einstein realized that the local effects of general relativity should also be seen everywhere, in the behavior of the entire universe. We've seen that he missed predicting the expanding universe. But his predictions that light is bent by mass have been confirmed by hundreds of cases where light from a distant galaxy is deflected, distorted, and even magnified by an intervening galaxy. The universe is a huge optics experiment based on gravity. Light bends whether the mass is visible or not. Based on this bending effect, astronomers have developed a method called *gravitational lensing*, which can "weigh" the mass of a galaxy or galaxy cluster by how much it bends light. Lensing has yielded evidence that something unseen—dark matter—is bending light, and a gravity calculation shows that it makes up most of the mass in the universe. The monks pepper me with questions, and I deal with them as well as I can. Then I'm saved by the call for tea time; I'm so tired I can barely stand. But the show's not over yet.

After the break there's an unexpected visitor.

"Ach, my feet, they ache! Himmel, why so many men in robes?

Ich bin verloren. I'm lost. I took the tram from Marktplatz across the middle bridge, but this is not Basel. Wo bin ich? Welches jahr haben wir? Ach, it's most upsetting when time and space get mixed up, but it seems to be happening to me a lot lately."

Even these otherworldly monks know the unmistakable visage of Einstein. The mask is a dead ringer. It bears his iconic crazed mane of hair, the sad eyes, and the lugubrious expression. From the inside, the mask is claustrophobic and uncomfortable. Air. I need air. I guess I asked for it. The summer sun is cooking our little building. I can barely see out through the mask's small eye holes, so I make my way gingerly to the front of the room. I'm hunched over to represent the great man in his dotage. The monks point and clap their hands. They hoot with laughter. I'm dressed in a tweed jacket and baggy pants. Sweat is pouring down my face and neck. My shirt is already soaked.

"Fragen?" I say. "Any questions?" says the translator.

They come thick and fast, and Karma has trouble keeping up. What's inside a black hole?

"Ja, it must be those woolly socks Elsa knits for me, I can never find them."

But nothing gets out, not even light. How does mass bend light?

"Gute frage, the photon's energy has equivalent mass so it is gravity, alles ist Schwerkraft."

What is the universe expanding into?

"Mein Gott, if I knew that I would have ten Nobel Prizes. Ganz im ernst, but seriously, my theory doesn't say; the space-time is curved so there is no need of an external medium."

Why was there a big bang?

"Ach du liebe, you are very clever monks, such questions you ask. I have no idea! Ich habe keine ahnung."

Einstein is real to the monks. They commit to the illusion. He spends twenty-five minutes answering questions, then waves and shuffles out the door. I reenter as myself a minute later and ask why the room is still buzzing with excitement. Then, having just impersonated such a famous personage, I can't resist telling a story that's probably apocryphal and has been elevated to the status of an urban legend.

Late in his life, Einstein was an iconic figure, the only scientist known to the general public. He liked to give public talks on relativity for "the masses," and preferred to travel to his speaking engagements in Princeton on a bone-shaker bicycle. His home institution, the Institute for Advanced Studies, felt this was beneath him and pressed him to make use of a chauffeur and limo. Einstein hated ceremony and artifice so he resisted, but they kept pressuring until he relented.

Einstein befriended the chauffeur and they enjoyed many chats on the way to and from speaking engagements. Months later, while they were driving, the chauffeur jokingly said, "You know, Professor Einstein, I've heard you give that talk so often I bet I could give it myself." Einstein got a twinkle in his eye and said, "Well, many people don't know what I look like, so why not!" Next time, they swapped clothes and Einstein drove the chauffeur to the talk. Einstein was shorter so had to turn up the cuffs on the chauffeur's uniform, and Einstein's tweed jacket rode up the chauffeur's arms, so they looked like an odd couple walking into the auditorium. However, the chauffeur had listened well and he gave a quite serviceable lecture. Einstein, for his part, relished sitting at the back of the room, and he goofed off reading comics and drinking soda.

At the end of the talk, there were questions, and the chauffeur batted away the first few relatively easily. Then a hand shot up in the front row. It was a young, intense man whose hair stood up like a brush and whose eyeglasses were thick like the bottom of a Coke bottle. "Professor Einstein, that's all very well, but what are the terms in the stress-energy-momentum tensor?" The chauffeur paused for a minute, then gave the man a withering look. "That's a very basic question." He gestured toward the back of the room. "In fact, it's so obvious that I'm going to have my chauffeur answer it."

General relativity is a shimmering monument to Einstein's genius. The great Indian astrophysicist Subramanyan Chandrasekhar asserted that it was "the most beautiful of all existing theories," and Einstein himself said, "Scarcely anyone who fully understands this theory can escape from its magic." This aesthetic sensibility echoes the famous words of John Keats at the end of the poem "Ode on a Grecian Urn,"

"Beauty is truth, truth beauty—that is all ye know on Earth, and all ye need to know."

The Story So Far

I'm losing all my toys. First it was the mystery ball, which I last saw one evening a week ago when a group of monks were huddled around it in the courtyard, interrogating it by moonlight. It has totemic power for them. Next it was the Zoob, which I keep in a plastic attaché case. I loaned it to Dawa and Jigme a few days ago and haven't seen it since. Now the Einstein mask. It gets passed around, appearing above monk robes, mugging for cameras.

A good night's sleep could save me from thinking about the lost toys, but I have insomnia tonight. There's no obvious reason. Sometimes my mind races. The worry ball rolls down the hill gathering flotsam and jetsam until sleep is elusive. I'm controlled by my monkey brain—a Buddhist might laugh.

The sky is just beginning to get pale. If I turn on a light I might wake Paul, so I join the monks for breakfast. They've already been up for hours and are loud and boisterous. There's jostling in the long maroon line for food, but it's not to get served, it's to *not* get served.

"You go first."

"No, you. I don't want to go first. You go."

"No, you."

This good-natured pantomime turns into a standoff as the food gets cold. To resolve the impasse, Thupten B steers me gently but firmly to the front of the line. Food at the workshops is better than monks get at their home monasteries. They pile their plates high with rice, dal, momos, and vegetables. I opt for one of the unleavened pancakes spread with butter and jam.

Geshe Nyima gives me a lesson on eating tsampa. Tsampa is the Tibetan staple: roast barley ground to the consistency of rough brown flour. Nyima adds salty butter tea and kneads it into a sticky paste. Four-year-olds would love this. He picks up a great wad of it and chews it between sips of tea. Across the table Gelek uses a different approach: he mixes the flour with less tea, meticulously rolls it into small balls, and pops them like M&M's. I experiment with this finger food. My approach must be too fastidious because many of the monks are smirking and smiling. The tsampa tastes to me like a mixture of beer and play dough.

Tsampa gives the monks enough hedonistic pleasure that I wonder mischievously if it's a substitute for sex. Karma says that in Tibet they favor meat that has rotted, from which they extract the maggots. They also let yak and goat cheese rot, remove the worms, and eat it. For long trips across the plateau, tsampa is packed into saddlebags as the perfect energy food for a long trip. Raw meat is cured with sweat by putting it on the horse's back, under the blanket and saddle. I scrutinize Karma for signs that he's pulling my leg, but detect none.

I join the teachers at their breakfast a little while later. The mood is somber. Another monk in Tibet set himself on fire as a protest. He died yesterday after three agonizing days in the hospital. The local school of music and dance down the road has canceled a performance they'd planned for the Dalai Lama's birthday; His Holiness decided a celebration wouldn't be appropriate. We talk about the fears of many Tibetans that their cause will suffer when he dies.

In the classroom, most of the monks are seated. I'm waiting for the stragglers to sit on their mats when I have a sudden pang of doubt.

How successful is this "East meets West" experiment? I worry that the connections I'm drawing between science and Buddhism are forced or false. At times, I believe that the dialogue between Buddhism and science is essential. One of my students, Ngawang Sherab, summarizes that optimism during one of our evening sessions: "Science has collected a vast amount of in-depth knowledge about a lot of things and is considered reliable and valuable in all parts of the society," he told me. "Likewise, Buddhism has gained tremendous knowledge about our mind and is a religion based on a philosophical system. Science knows about our physical world and Buddhism knows about our inner world, and if they come together they won't just complement each other but also provide us with a whole or complete picture."

On other days, I'm not so sure. Buddhism talks of many worlds or realms of existence, but these ideas are metaphorical. They aren't directly related to the exoplanets being discovered by astronomers or to the multiverse concept in cosmology. Interdependence is a core Buddhist idea, but it's both broader and less well defined than the forces and fields that are the core of physics. *What the Bleep Do We Know?!* was one of the few movies I've ever walked out on. It drew a New Age connection between quantum physics and consciousness, and it also annoyed many of my colleagues. In bridging Buddhism and science, am I doing a disservice to both? Am I guilty of the same intellectual sloppiness? Are my bridges poor substitutes for formal preparation in math and physics? Doubt gnaws at me. But expectant faces are tilted up at me, too. So I begin the class.

We pick up the story of cosmic evolution. We started with conditions of unimaginable density and temperature, and then journeyed to a time

MATTER DISTRIBUTION GOES FROM SMOOTH TO LUMPY.

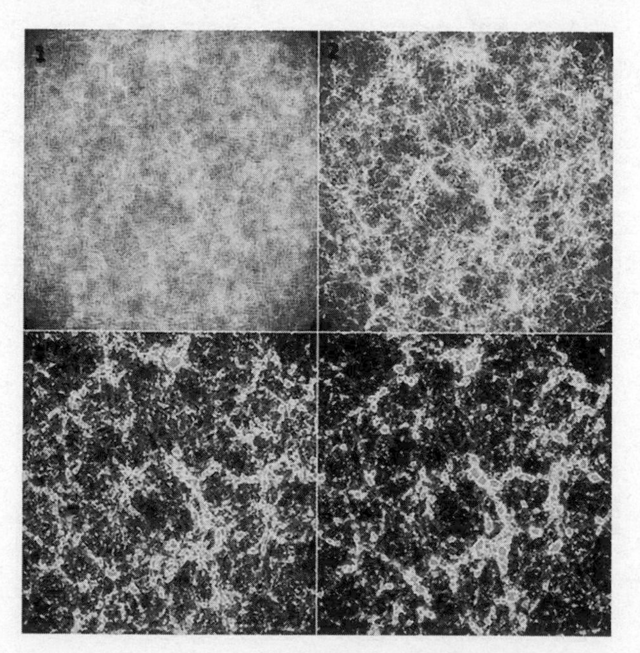

Figure 5.7. From smooth initial conditions, the single force of gravity sculpts structures on scales of tens of millions of light-years (in this supercomputer simulation, time flows from top left to bottom right).

four hundred thousand years after the big bang, when the universe became transparent. The next big event was the formation of galaxies, which arose out of expanding, congealing gas. Between the moment of transparency and the rise of galaxies, about 20 million years after the big bang, the universe was a comfy twenty-seven degrees Centigrade (eighty degrees Fahrenheit), or room temperature. It was also cozy—a hundred times smaller and a million times denser than it is now. Not that we could have survived then. There was only hydrogen and helium to breathe, and the density was a lung-busting, near-perfect vacuum. Nonetheless, it's interesting to try and imagine.

With that thought, I deviate from my lesson plan and extend the visualization experiment of the previous week. We will take ourselves to the largest scales humanly imaginable. Then we will zoom back in and go to the smallest imaginable scales. We're about to take a tour of all physical reality, but I don't let on. I ask for their trust. "Relax," I begin. "Close your eyes." The room gradually stills, and gets so quiet you could hear a momo drop. The monks are accustomed to prayer and meditation. They quickly settle into relaxed but alert states. Wind fluffs the curtains. We hear the distant rhythmic clack of road workers breaking stones. My monkey mind wanders. I mentally pinch myself to regain focus.

"Keep your attention inside," I say. Our journey goes like this:

> Listen to your breathing and lock onto that rhythm. Good. Now slowly release your attention so that it floats up and out of the room. You move swiftly upward. Look down but don't be distracted by the patchwork quilt of houses and fields; you're going on a long journey through time and space. Quickly you traverse the starscapes of the Milky Way, and its perfect symmetric spiral form recedes from view. There are galaxies all around and you marvel at the way each one differs slightly from the others, and how each fades smoothly and imperceptibly into the black of night.
>
> Your path takes you directly toward a huge elliptical galaxy, and it grows until it fills your vision. You flinch and

brace for an impact, but there is none; stars whip by you soundlessly on all sides, and for a moment you are within it—like being surrounded by a swarm of yellow bees. Then it rapidly recedes behind you.

You relax and start to enjoy the journey. Galaxies are above, below, and all around you, like fireflies. When one passes by closely you see that's it made of myriad stars. You travel with no destination in mind, the view is the same in all directions, but gradually you notice that the galaxies are closer together and it's warmer than when you started. Many of the galaxies seem to be interacting, tossing great arcs of stars into space like seeds sown by a farmer. The galaxies are crackling with flashes of light as massive stars within them detonate. The view is changing faster and faster.

Suddenly, it's dark, and getting very warm. The darkness persists for a while, then turns into a dull red glow. Then the color shifts through the rainbow until it's a vivid blue-white light, so intense it makes you catch your breath. You feel a momentary wave of panic, but then relax into the experience. You are freed from your body. Nothing can hurt you.

Now there's a crescendo. The blinding whiteness resolves into a chaos of particles moving so swiftly they are a blur. You sense there is heat and pressure beyond imagining around you, but you feel nothing. You lose your sense of moving through space. You have no separation from the space around you. The universe is the container, and yet you are the container. You shrink to a dot. The dot is nothing, yet it contains everything.

But as you shrink, it's like a spring is being pulled taut. Just when it feels like you will be obliterated, you release your concentration. The spring pulls you back and away. You don't resist, and it pulls back through the events you've witnessed, through white hot space fading to dark, the heat ebbing, and galaxies crystallizing from the void. With its last ounce of tension, the spring pulls you back into the Milky

Way, toward Earth, and into your body. You settle and breathe quietly until you hear the heat bugs outside and the muffled thud of your own heartbeat.

You are back in the room and in your body. Nothing has changed. It feels like no time has passed. You release your attention. It expands like a bubble. As it does your focus of concentration shrinks within it. It becomes a vital energy force, a *qi*, and it settles deep within you to a place in your abdomen. You let your attention expand, and the place of focus within becomes smaller and smaller. It shrinks within tissue in your chest to the level of a single cell, and on until the twisting ladder of DNA is apparent.

You are going down farther now. Atoms are becoming visible, their electron clouds shimmering like motes of dust caught in sunlight. You start to see the nucleus of the atom. Quarks are darting in and out of the nucleus in a blur. Then it's dark. Your focus keeps shrinking, but there's no structure and nothing to see. At the limit of your journey your attention shrinks to a dot. It has no size and no dimension—nothing with the possibility of everything. The darkness resolves into a bright lattice of waves and oscillations, like the surface of a moonlit ocean. You contemplate this alien world. Now you release your focus and it swells back into your body.

As we open our eyes, I summarize the journey. Our travel outward and our travel inward have met at the same point. We experienced the way nothing created everything. The void beyond and the void within are one. It's a closed circle, the snake that eats its tail. Ouroboros.

The monks' sense of play is infectious. I've come down with a healthy case of playfulness. During the first week my topics and learning goals were neatly arranged in a spreadsheet. I kept a close eye on the time to avoid getting behind. We did a lot of hands-on activities, but I resisted letting them spill out of their assigned time slots. Last week I got comfortable with the cadence of translation and had enough confidence in my students to fit the lectures around group work and discussion.

Now that I'm into the third and last week I want to throw away the playbook and improvise.

So why not videos and rock music I've smuggled across international borders? And snacks. Monks do not live by momos alone. On a trip to Dharamsala I've stocked up on Smarties—the British equivalent of M&Ms—and Bombay mix. We hand around the sweet and spicy treats and pull the curtains to darken the room. I've given Tenzin the lyrics for the music we'll hear, and he's passed around a written translation.

First up is Pink Floyd, whose song, "Learning to Fly," accompanies a supercomputer simulation on my computer screen. We watch the formation of large scale structure in the universe. As plangent guitar chords soar, galaxies congeal from a heaving sea of dark matter. The monks read the lyrics: "Into the distance, a river of black stretched to the point of no turning back." The music is propulsive, hypnotic. I'm foisting 1980s musical sensibilities on my monastic wards, but they seem to be engaged. Some sway to the music.

Next I play a scene from Monty Python's movie *Meaning of Life* where Eric Idle, wearing a pink tuxedo, emerges from a refrigerator, to show Terry Jones, who's in drag as a housewife, the wonders of the universe. Their "Insignificance Song" contains a lot of astronomical information. And it's very catchy. By the end, many of the monks are humming the chorus. Few people could disagree with sentiment of the closing lines: "And pray that there's intelligent life somewhere up in space, because there's bugger all down here on Earth." I wonder if this is the first time Python and Floyd have been translated into Tibetan.

I must stop being naughty. In the *Ghitassara Suta*, the Buddha listed the dangers of singing and chanting. First, you get attached to the sound. Second, other people get attached to the sound. Third, it can annoy the neighbors. Fourth, the concentration of those who do not like the sound can be destroyed. Fifth, later generations might copy it. I hope my contextual use of popular music is in keeping with the Middle Way.

Our room is still dark when I turn off the digital entertainment. Through a gap in the curtains, the Sun hangs low like an orb of blood

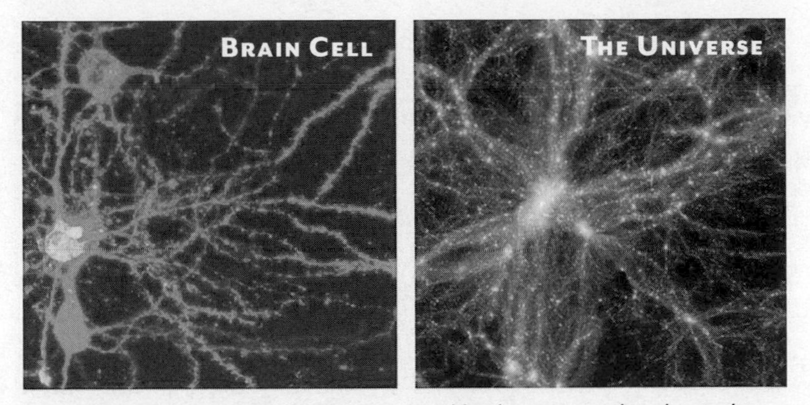

Sometimes the structures in physics and biology are strikingly similar, and can be described by similar mathematical forms.

Figure 5.8. Cosmic structures are so complex and subtle it can seem that they must be controlled by intelligence rather than blind forces. Similar types of structures are seen in the wiring of the human brain.

orange. It's time to wrap up for the day with another theatrical event, the last but not the least. The curtains are pulled open wide.

Growing up in England, theater was part of our family activities. I recall the folk tradition of the "Mummer's Play." In medieval times, troupes of actors traveled the countryside. They went village to village putting on outdoor plays. With colorful costumes and rudimentary props, they acted out stories passed down for centuries: The Plague; Fool's Lament; Hunting the Hare. Over the past several days, we've adapted a similar set of simple stories. The monks have enacted the entire cosmic drama. If we had a playbill, it would look like this:

Act I. Forces of Nature: A Masquerade, or, "How Everything Came from Nothing and Four Colorful Characters Emerged."

Act II. Monk Gravity: A Minuet with Ten Thousand Dancers, or "How the Universe Changed from Smooth to Lumpy."

Now we'll perform Act III and then take our bows. We call this final act: Earth Begins: A Waltz, or, "How Angular Momentum Made Galaxies, the Sun, and the Earth."

I ask the monks to stand. I remind them of our earlier scenario when they all pulled on each other, illustrating how a finite Newtonian universe is unstable, prone to collapse. We'll all pull on each other again, but will add a twist—or rather, a rotation. Gas clouds in the universe slightly rotate. This speeds up when gravity collapses a cloud on the way to forming galaxies, star clusters, or individual stars. To demonstrate, I bring forward an old swivel chair, modified to be like a turntable.

I scan the room. I need a strong monk with long arms. Gelek! He sits cross-legged on our turntable. I extend his arms out sideways and put a brick in each hand. Since India is littered with building sites, bricks are easy to come by. I spin him until he's going at a fair clip, about one revolution every two seconds. Everyone crowds in to watch. "Now pull your hands in toward your chest." As he does, his spin rate increases. He becomes a blur. After I drag him to a halt, he gets up and staggers to the side. The troupe hoots and chortles.

We've just seen conservation of angular momentum. It's yet another foundation stone of physics. When a rotating system's mass becomes more compact, the rate of rotation increases, which is the story of any cosmic gas cloud that collapses. Starting from a diffuse state with slow rotation, it ends up in a more compact configuration rotating faster. Gelek has shown us the principle for an individual monk, so we will now act it out with a "cloud" of monks. We practice once. Then the curtain rises on the final act: Earth Begins.

The monk cloud fills most of the room. It shuffles slowly clockwise. This is a nod to Buddhism's directional preference, although the universe seems not to be mindful of either clocks or religion. As gravity pulls the monks closer together, they move faster until they're almost running. Then, by prior agreement, Geshe Nyima takes the center position, his girth indicating that he's the Sun. Eight other monks settle into respectful circles around him in a reasonable facsimile of Kepler's elliptical orbits. I approach the third monk out, Jigme. I hand him a small, blue and green metal sphere, familiar from our scale-model activity in the first week. He cradles it carefully and tilts his head to look at it. It's something special: our home.

The Meaning of Life

As I awake, the first thing I hear is a low, wavering hum, like bumblebees in flight: monks chanting. The bass line is overlaid with a treble of birdsong. I open my eyes. Pale blue-gray light seeps in through the cotton drape. I've no idea what day it is. My normal life has slipped away until nothing is left but a carapace, the bright shell that survives the withered-away insect.

The chanting below our window is as recognizable as the birds at dawn. Both greet the day with an expression from the core of their being. To chant or sing the world into existence every day is a minor miracle. Other small miracles abound. Paul is feeling much better. He has taken an intellectual journey, having devoured all six books that he brought along to read. His curiosity has turned to the box of Zoob, which I've miraculously found and retrieved from the monks. Before my very eyes, Paul becomes the "Prince of Zoob."

With the Zoob back in our bedroom, he's quickly drawn in and he pulls me along in his wake. Father, son, and toy. It seems just right for the edge of the Himalayas. Paul explores the full range of basic modules. Then he embarks on the construction of alternative worlds. In just one evening he comes up with indestructible cubes and compressible sheets, crisp lattices and amorphous tangles, nested structures and elegant knots. Another evening he creates all the basic components of miniature machines—levers, hinges, axles, gears, and ratchets. With a glint in his eye he says, "With enough Zoob I could make anything." My chest swells. I've fathered a brilliant but mad scientist.

I watch Paul's facility. I admire his nimble fingers. The room is quiet except for the click and snap of plastic, and the occasional loud curse. It's quite a large box of Zoob, I realize. Soon every surface in our small room is covered with rainbow constructions.

Then we're on the road again, on one of our routine trips to Dharamsala. We gawk at tourists and stock up on cookies and books. At the small post office we buy postcard stamps but they have no stick. Is this so people cement them down so hard that they can't be reused? We go off to buy some glue, and the clerk postmarks the cards as we watch. We're tempted to conclude that postal fraud is an issue. Yet India, for the most part, is a scrupulously honest country. We've even seen the Indian version of America's "Honest Abe" story, in which shopkeeper Abraham Lincoln travels far to return some money. Several times I've had extra change returned when, clearly, I didn't know the price or denominations. One shopkeeper even ran down the street after me. More evidence of public probity: during our narrow-gauge-train ride, Paul and I forgot our umbrellas, and three hours and ten stops later, we found them where we left them.

I have plenty of mental sustenance but feel the need for a physical challenge. Running is still brutal. Most days it's eighty degrees with eighty percent humidity. On this particular day, I'm on a ridge, so I can go up or down. The winding road up passes through tiny Sarah, where I've become a known quantity. Kids and shopkeepers wave. When I pass the dank and grimy one-room schoolhouse, I hear the hectoring voice of the teacher. It sounds more like a tongue-lashing than a lesson. I cruise by rows of brick buildings with one wall open to the road and everything inside revealed, like doll houses. Inside, two-stroke engines run mills to grind aggregate into cement, or wheat into flour, or barley into tsampa. At the edge of town I see a flatbed truck with a tiny temple on it. Four people are crammed inside, all wearing identical gold and emerald robes and headdresses. It's a Porta-Temple. For minor festivals and feast days it can't be beat.

There are dogs everywhere in the villages, mangy and skeletal, lying by the side of the street. A dog's life is hard, but a cat's life is even harder. Just after we arrived, the cook's cat had a litter of eight. Only one had survived a week later; the rest were killed by dogs. I've seen a mongoose scurry across the road and vultures arcing high in the sky. Venturing off-road, the tangled undergrowth is alive with birds, butterflies, and beetles. As I jog along I stay alert since northern India is also home to some of the most venomous snakes in the world: kraits, vipers, and cobras.

Following a downward path, the road leaves the forest and crosses a river valley that eventually feeds into the mighty Ganges and the sea. I take this route because the path up the hillside looks dark and ominous. I come upon a tent encampment on a scrap of land between the road and the river. Families are living there with no creature comforts. Two little girls in rag dresses dash out with hands open. I feel bad that I've nothing to give them.

On the way back I realize I've miscalculated. Rain is surging back down the valley. Soon I'm running through dark sheets of water. The rocky path I followed has turned into an ankle-deep torrent. One good thing about water: there's no category beyond soaked. As Paul says to me, I "got pwned by it."

Back at the College for Higher Tibetan Studies, my circle of attention has shrunk in a satisfying way. The world beyond this dilapidated campus is impressionistic. I'm aware that a clamor of e-mail awaits me, along with the buzz of social media and the twenty-four-hour news cycle. But I don't care. My sleep is governed by the ebb and flow of sunlight. My body is gently cleansed by rice and dal. I'm stuck with my monkey mind. The self-doubt and insecurity engine still runs hot on occasion. And I'm still prone to overinvestment in knowing stuff and being right, both as a scientist and in my personal life. But in idle moments creative thoughts are appearing with increasing frequency. Perhaps I can stay here? It's a tempting—and creative—thought.

For two and a half weeks, Paul and I have lived together in a ten-by-twelve-foot room. We're getting on really well. My son is a very cool kid, with a restless mind and an unsullied sweetness I hope he never loses. We talk about astronomy and physics late into the night. He taxes me with high-level questions about black holes and Hawking radiation and horizons in cosmology.

He's particularly interested in the big bang and the multiverse—topics that also intrigue the monks. While the vastness of the universe gives me a creeping sense of existential angst, he maintains an optimistic sense of meaning to it all, in our lives, and also out there in the jungles and galaxies. He's not worried about nothingness. At this moment, we've been talking for several hours. I'm fending off sleep as we discuss the potential of the multiverse. He says, "Everything that's possible not only can happen but is happening right now as we say this." He sounds satisfied. My dwindling circle of consciousness shrinks to a dot.

Life on Earth

To realize how life grips this planet like a fever, I've set the monks a challenge. Go outside at the break and try and find a patch of land or dirt the size of your hand, any place at all, where there's no life. I've brought a magnifying glass with me but don't think we'll need it. With their naked eyes, I'm sure the monks won't be able to escape mounds of evidence for our final topic: life on the Earth (and elsewhere).

They spill out of the building and fan out across the grounds. Some head out of the gate or beyond the wall that rings the monastery. They avoid the grassy areas and head for wasteland. Approaching one group, I see they're staring at handfuls of soil. It's crawling with worms and bugs of many kinds. Another group has found a pile of builder's sand. That, too, has insects crawling in it. Even a tiny white flower gets nourishment there. Finally, Thupten hands me a rock and gives me a triumphant look. It's an igneous rock, hard and granitic.

"Maybe," I say.

I bring out the magnifying glass. We inspect the rock and Thupten sighs. It's covered with tiny yellowish specks of lichen. Lichen isn't just alive; it's a complex organism in symbiotic relationships with algae and bacteria.

The fecundity of the planet is surprising even to biologists. The soil in this area of India is poor but seething with life. In addition to easily visible worms and insects, there are mites and nematodes no bigger than the head of a pin. Soil harbors yeast and fungi and protozoa and algae and bacteria and viruses. There are 50 billion microbial organisms in a tablespoon of dirt.

It's the same story in the oceans. Take any liter from the ocean and it might not have any fish in it but it would have an average of thirty-eight thousand microbial species. A teaspoon of sea water has more DNA than the human genome, all in the form of invisible, microscopic organisms. It's an overwhelmingly microbial world, which is true of us as well. One project to catalogue microbes in the human body counted ten for every human cell. However, they're small, so we only carry around a few pounds of bacteria. It makes me wonder: Are they the tenants or are we?

Now that we've primed the pump with observation, we can address an important scientific question: How did Earth get this way, and is there anything special about the story of life on this planet?

We start by considering the difference between us, air, dirt, and the Sun. We're made of about two-thirds water, which is hydrogen and oxygen in the ratio of 2 to 1. After 61 percent hydrogen and 26 percent oxygen, there's 11 percent carbon and 2 percent nitrogen. Everything else is under 1 percent. Air is 78 percent nitrogen and 21 percent oxygen, and everything else is under 1 percent. Dirt varies in chemical composition from place to place, though a typical breakdown would be 45 percent oxygen, 30 percent silicon, 8 percent aluminum, 5 percent iron, and a few percent each of calcium, sodium, and potassium.

Compared to this rich brew of hefty elements, the Sun is strikingly simple. It's overwhelmingly made of the two lightest stable elements: 91 percent hydrogen, 9 percent helium, and less than a percent of all others—carbon and oxygen being the most common trace ingredients. The same is true of all the stars in the universe. We are "star stuff" after the carbon, nitrogen, and oxygen have been concentrated to be hundreds of times more abundant.

The story of life starts with the first stars in the universe. Biologists think life has a minimum of three requirements—carbon, energy, and water. The primordial universe contained only hydrogen, the simplest element, and the helium that was fused in the first few minutes after the big bang. Those two ingredients account for 99.9 percent of the normal matter (dark "matter" being another matter entirely). The rest is the result of stellar alchemy and generation upon generation of stars living and dying.

Our story is intimately connected with the stars. We're both literally and metaphorically stardust. The engine is fusion in the cores of stars, where the temperature and pressure are enough for atomic nuclei to overcome their mutual electrical repulsion and fuse to make a heavier element. The higher the star's mass, the higher up the fusion chain it can ascend, because more mass leads to a higher core temperature. In low-mass stars like the Sun, a little bit of carbon is created and that's all. In higher-mass stars, elements up to iron are forged. As part of their evolution, massive stars "burp" gas into space that's enriched with heavy elements. The most massive stars "barf" gas into space, and the cataclysmic explosion that happens when they die can make

COSMIC ABUNDANCE

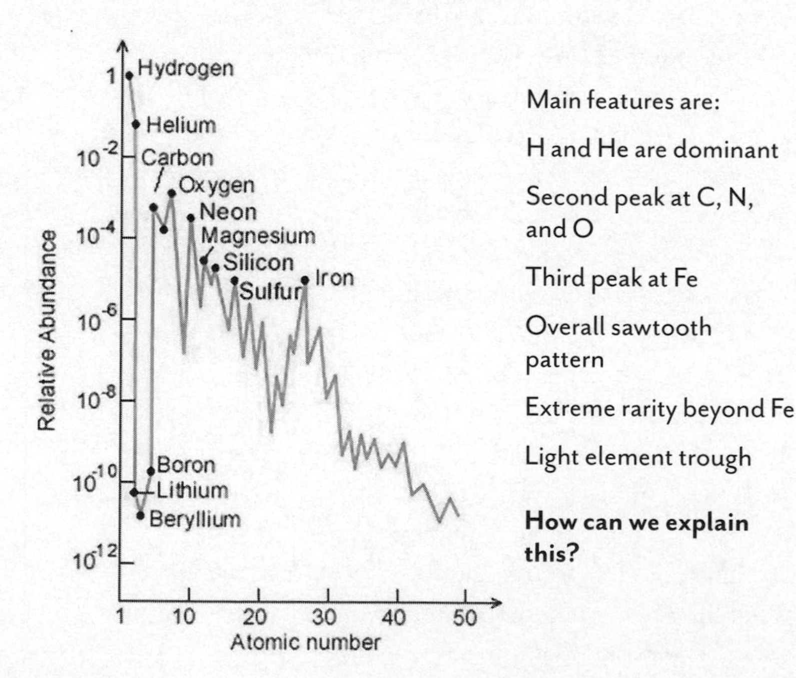

Figure 6.1. The story of life is a story of stars. The heavy elements from carbon and on up the periodic table were generated in the cores of stars or the blast waves of dying stars and ejected into interstellar space throughout the universe.

the heaviest elements. In this way, gold, silver, platinum, uranium, and other precious and radioactive elements are created and flung into the cosmos.

When the first generation of stars formed, a few hundred million years after the big bang, there were no planets and no people because there was nothing from which to make them. Over time, stars have created the heavier elements and returned some fraction of that gas into the space between stars, where it can become part of the next generation of stars. The cycle continues. Stars form with successively heavier elements mixed into their gas and so are able to make planets. Carbon harbored by the Sun will one day be returned to the interstellar medium where it might become part of biology elsewhere in the far future. In a sense, the universe has been getting more hospitable to planets and life since the big bang.

It doesn't take much. Adding up the rocky inner planets and the rocky cores of the gas giant planets, and tossing in the asteroids, gives a total of about twenty Earth masses. That's about 0.01 percent of the mass of the Sun. When a star forms, just a tiny bit of grit is left over, but like the grit that grows a pearl, it's interesting to see the result.

How boring would it be if stars had never formed? Let's find out.

Some of the monks have been lobbying for the Return of Zoob. Having finally got the Zoob back, and with Paul having exhausted its possibilities for the moment, I'm not about to disappoint them. I extracted all the yellow pieces and distribute them among the monks, who are working in groups. Yellow Zoob has a ball at either end, and a notch set into the rod that connects them. "There's your primordial universe," I tell them. Go ahead and create! It doesn't take long for them to run out of options. All you can do with yellow is connect the notches crossways, making a little X. Balls can't connect to balls, so that's it. In this model of chemistry with one atom, you can only make one simple molecule. One and done. It's like the universe before the first stars, containing hydrogen and helium. Two hydrogen atoms can pair to make the simplest possible molecule, but helium is chemically aloof so no additional molecules are possible.

I walk around the room and distribute star barf, or at least something that will stand in for carbon. Each group gets a pile of green Zoobs to go with their yellow Zoobs. Green Zoob has a jaw at either end. Now the game is on. Green pieces can connect end to end to make long linear chains, and the jaws can grab the middle of any piece to form a right-angled connection. Playing just with green Zoob, the monks are soon forming lattices and cubes and sheets, and when the sheets get large enough they curve them into tubes. There are sidelong glances between groups to see who's making the best stuff.

Green Zoob is a great analog for carbon, which easily forms long chains, and is found in nature in forms as diverse as diamonds, graphite, and buckyballs. Already, some groups are combining yellow and green pieces into Byzantine structures, and the buzz in the room is palpable. The ball at each end of a yellow piece is pitted, and the texture allows the jaw of a green piece to grab it at a lot of different angles. Suddenly, the possibilities explode. Shapes can curve and undulate and interpenetrate. The monks make amorphous structures, without any obvious pattern. They've jumped headlong into organic chemistry without any training.

We pick up our story on the primitive Earth. I set the scene for them. Nine billion years after the big bang, in the suburbs of an otherwise undistinguished spiral galaxy, a star dies and triggers the collapse of a large and diffuse gas cloud. Within the cloud a new star ignites, and the cooler and heavier elements condense into tiny particles. In 50 million years, less than the blink of a cosmic eye, the particles aggregate into rocky snowflakes, then dust bunnies, then rocks, then mountains, then planetary embryos the size of large cities, then a set of eight rocky planet cores. The planet we call Earth ends up in what astronomers have dubbed the "Goldilocks zone" of the solar system, where the temperature at its surface is "just right" for liquid water.

On the newly minted Earth, however, Goldilocks's desire to have her porridge temperature just right would have been seriously thwarted. The young Sun was feisty, hot, and variable. There was no breathable atmosphere. Earth was molten with no rock to stand on. There were no oceans because water clung to the planet as a shroud of steam. Huge meteors crashed down every few weeks. If Goldilocks survived all of this, she'd eventually see solid ground and oceans form. Life probably got started soon after that.

- **Water** is special (and it is a common molecule throughout the universe).

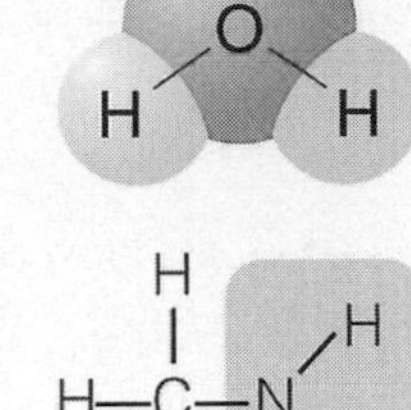

- **Carbon** is special (and it's the second most abundant element made in stars).

- With a versatile chemical tool kit, the possibilities of information storage and biological function are almost infinite. Alternate biologies are definitely possible. We just have to look for them.

Figure 6.2. Biology on the Earth—and presumably beyond the Earth, if such biology exists—uses the cosmically abundant ingredients of liquid water as the medium for chemical interactions and carbon for the information storage.

Nobody knows where it happened. It could have been near hydrothermal vents on the sea floor, or inside rock in the crust, or near the edge of a shallow lake. Evidence of single-celled organisms has been found in 3.7-billion-year-old rocks, with hints of life in rocks that are 3.9 to 4 billion years old.

It's pure speculation how life first started. Nobody was there to witness the event. Earth has had so much volcanism and erosion that it's very hard to find unaltered rocks from that long ago. Think of the difference between simple molecules dissolved in water and cells. Even the simplest cell is a sophisticated biochemical factory, with a complex network of thousands of chemical reactions, where function and form are stored and transmitted in the double helix of DNA. Scientists have simulated aspects of early evolution. They've shown that simple molecules will interact and become more complex if there's energy available. They've shown that fatty molecules naturally form little containers that repel water and concentrate reactions. They've shown that the rudiments of an information-storing molecule can develop. But nobody has ever made life from scratch in the way we think happened on the Earth. If only we had Goldilocks as a witness.

How do we know the origin of the first life wasn't a fluke? This is my rhetorical questions for the monks. "We don't know," is the answer. I remind them that this is historical science. We may never know exactly how it happened. But there's no reason to appeal to divine intervention or supernatural causes. Scientists are confident that biology emerged naturally from chemistry.

Biology exhibits both unity and interdependence, and in this sense it recapitulates Buddhism. We're all one thing. From fungi to elephants, all life on Earth derives from a single biological experiment. It shares the same genetic code. An alphabet of four chemical bases (A, T, C, and G—adenine, thymine, cytosine, and guanine) forms the "letters" of the code. They combine in groups of three for the "words" that specify twenty different amino acids used by living organisms. We can think of a gene as a larger unit of genetic information like a "sentence," specifying an attribute of the organism. The complete genetic information of a human is 3 billion bases in a sequence that reads like the book of life.

My biology colleague, Gail Burd, mentioned earlier that the genetics revolution is generally unknown at the edge of the Himalayas. The monks are clearly fascinated, and they think it has real meaning for their lives. I show them a picture of the classic double helix of DNA, with the bases color coded and the rules for bases connecting shown: A pairs only with T, and C pairs only with G. It looks like a twisted ladder with base pairs forming the rungs. The double helix means that when one side of the code is read we can deduce the sequence on the complementary strand, which is the key to how DNA stores and transmits information.

"Did the genetic code have to be this way?" Jigme asks. It's a very perceptive question. Probably not, I say, since presumably the coding could have worked in several different ways. Science can make progress by considering counterfactuals—plausible ideas that don't represent the path nature actually took. It suddenly occurs to me that we can explore this idea with the Zoob. First I ask the groups to make long strands for the outer rails of the ladder. Then I let them experiment with the number of ways they can connect the rails using the four Zoob pieces that have jaws: red, blue, green, and gray. For an educator, this kind of unstructured activity is fascinating to watch. I've no idea what they'll come up with.

After ten minutes the answer is clear. Zoob can store more information than the actual genetic code because each of the four colors can pair up in more than two ways. I try to explain why, and it's not easy: "The efficiency of coding comes at the expense of robust transcription and with more ambiguity in the sequences." I hope Tenzin got it. We talk about whether the genetic code has to be perfect. Posing as real research biologists, and doing experiments on hypothetical biology, the monks correctly conclude that the code just has to be "good enough." It's another kind of Goldilocks zone.

After a tea break we do some DNA extraction. Genetics lab equipment is impossible to find in Himachal Pradesh—my biggest challenge is getting hold of some strawberries. After trips to the nearest three villages, I'm ready to give up; only locally grown fruit is available. Wryly, I realize my Western impatience in assuming I can get anything I want at any time of the year in a market. One general-purpose shop in Dharamsala has a few dusty cans of fruit, including strawberries, and they'll do. The other equipment I need is mundane: plastic bags, ice, salt, alcohol, shampoo, and small wooden satay skewers. I've brought test tubes, cheesecloth, and a white lab coat. Ceremoniously I appoint Jigme as the lab boss and ease the white coat over his red robes. He glows with pride and laughs maniacally, a mad monk scientist.

Working in groups, the monks mash the strawberries into a pulp in plastic bags, along with shampoo, a pinch of salt, and a little water. Strawberries work well because they have large cells. The cell walls are broken by the mashing and further dissolved by detergent in the shampoo. Each group then filters the pink glop through a sheet of cheesecloth to get rid of as much of the cell containers as possible. I see that most of them aren't at all squeamish about getting messy. "Dawa, don't eat the experiment!" I glare at him with mock severity.

They put a little strawberry goo into the bottom of a test tube and then conduct the next step very carefully; they add some alcohol, which has been cooling in an ice bath. Walking around the room I can see many but not all the groups have milky material precipitating at the alcohol-water boundary. Taking the wooden skewer they reach down to the interface and extract a little material. There's great excitement in Gelek's group. He's holding the skewer up. The others are huddled close, their faces inches from the tip. I can clearly see clumpy but linear strands half a centimeter long. It's an aggregate of many strands, along with bits of cell wall, but there's no mistaking their success. DNA! Each member of the group gets a NASA sticker, causing a few envious glances from around the room.

I show a graph of the modern "tree of life" based on the overlap of the base pair sequence. Evolution is traced using the gradual deviation of DNA or RNA caused by mutation or genetic mixing. Most of the tree is taken up by microbes because most of the biological diversity on Earth is in microbial species. In terms of DNA, animals are just one branch of the tree and humans are a tiny twiglet on the primate twig. With all of terrestrial biology being just one continuous experiment, the deviations among species aren't large. We share 99 percent of our DNA with chimpanzees, 87 percent with dogs, 64 percent with moths, 60 percent with strawberries, and 55 percent with yeast. Look at that pink goo you made, I say. "It's half you!"

Thupten's hand shoots up. "If we share half of our DNA with fruit and get half our DNA from each parent, is a piece of fruit the same as one of our parents?"

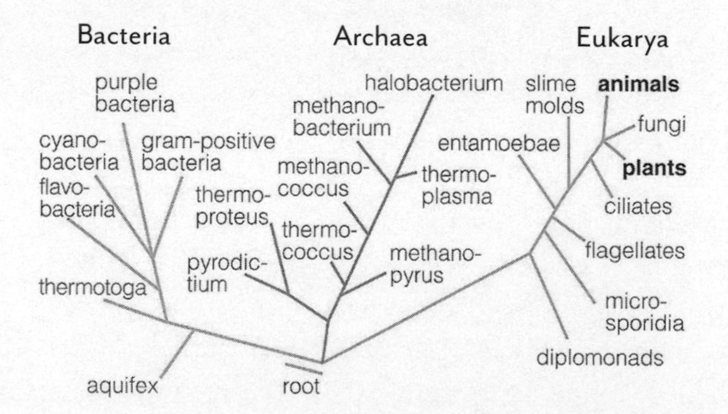

Figure 6.3. The modern tree of life evolves upwards in this diagram based on gradual variation of the genetic material since a primordial "last common ancestor." Humans are a tiny twig on the very small genetic branch labeled "animals."

The room erupts in laughter. It takes me a bit longer to get the joke as I wait for Karma to translate. "Great question!" The overlap of DNA is at the level of base pairs, the letters of the genetic alphabet or the lowest level of organization. All species are based on cells, and all cells need to organize chemicals and store energy and get rid of waste products, so we'd expect DNA to have substantial similarities. Attributes of the species, like having legs or a backbone, or getting energy by sunlight or chemical means, are specified by genes. For humans, those genes say whether we have black hair, or like coffee, or can jump high, or are prone to getting cancer. Visible differences like skin and hair color and the shape of our noses and eyes are controlled by very few genes, a handful out of twenty thousand. The distinctions that trigger racial hatred or cause us to declare a group as "other" are based on minor genetic variations.

I walk back down the aisle and pull Dawa to his feet and lead him up to the front of the classroom. I wrap my arm around this small, sloe-eyed, brown man, who barely comes up to my shoulder. "You're my brother," I say. He beams at me. And despite the gap of culture and language, it's true. Genetics makes a mockery of racism and our ill-guided attempts to define people by their appearance.

Look at your neighbor. The monks turn to look at each other. They're all of the Tibetan "tribe," and while they do look different from each other, they share skin and hair color and the general shape of their facial features. The genetic difference between any two of you, I tell them, is bigger than the genetic difference between you and a Greek farmer. The same goes for you and an Inuit fisherman, or you and a Bantu tribesman. It's a small world after all.

Dreams of Other Worlds

That evening I walk to dinner as the sun is setting and hear shouts and activity coming from behind the temple. Rounding the corner I see a patch of cracked asphalt with a basketball hoop at one end. A dozen of the monks are playing a pickup game—which I thought was forbidden—in the fading light. It seems that the rules are looser when the monks aren't in their home monasteries, another reason they enjoy coming to these workshops. They wave me over and I join in.

The teams are distinguished only by the use of prayer beads wrapped around the wrists of one team. I imagine it will be a genteel game but I'm wrong. Many of the monks are decent shots. They all know how to use their elbows. They hustle the ball up and down the court with sandals flapping and maroon robes swirling. Gelek uses his height to maximum advantage, and Thupten hand-checks me mercilessly. We play until dusk ends and we can't see the basket. Monks are not generally athletic but a few have definitely got game.

The next morning we put everything we've learned about the universe and the Earth together with a timeline activity. I'm hoping to leverage the monks' great debating skills.

The activity is simple. I've prepared thirty-four color pictures, a mixture of deep space images and events on the Earth. The goal is to lay them out on the floor in chronological order from the big bang to now. Each group of monks has a set. They'll work on their own, and then the groups will move around the room and critique each other's work. I've allotted about twenty minutes for this activity. Some of the items will be easy to identify and place, but I'm guessing several will be tricky for them to get right. I tell them if they think two or more events happened at the same time, they can place them side by side.

For the monks to get it right, they will have to line up the images in this order, beginning 13.8 billion years ago:

Equal numbers of particles and antiparticles
Hydrogen becomes helium in intense heat
Cosmic microwave background radiation
Large galaxies assemble from little galaxies
Quasi-stellar objects (QSOs)
Elliptical galaxies
Gold nuggets
Clusters of galaxies
The Sun, then Earth and Mars, then the Moon

Earth's oceans
Stromatolite (microbial) colonies
Trilobites, then starfish, then fish
Ferns, insects
Lizards, then dinosaurs
Small mammals, then octopi, then horses
Chimps, then early humans
Sailing boats, then the Pyramids, then the Great Wall of China
The first telescope, then an early car, then an early airplane
The Space Shuttle, then an early cell phone.

The groups quickly lay out images on the floor. The discussion grows animated. I move from group to group. Tenzin gives me a taste of what's going on. Some placements must be exact; you'd need to know that the Moon was formed slightly after the Earth by an impact, and that fish evolved before life moved onto land and gave rise to plants, or that the Space Shuttle preceded the first cell phone by a few years. Other parts of the ordering are judgment calls. Gold on Earth was created in stars that died before the Sun was born, but whether most gold was synthesized before or after large red galaxies isn't clear. After twenty minutes, I signal time and walk around the room. Most groups have most of the images in the right order. But no two groups have exactly same order, and there are some glaring misplacements.

Now I declare a mutual inspection phase, and the monks jump at the chance to see what other groups have done well—or poorly.

One group that finished quickly and with apparent confidence crosses over to another group that's still sitting on the floor, grappling with its timeline. A vigorous debate ensues between members of the two camps. Monk debate is nothing like a formal debate in an American high school adhering to Robert's Rules of Order, or the often sterile political debates seen in Washington. It's an intriguing mixture of common sense, sharp critique, esoteric ideas, and physical dynamism.

We're about to have a monk melee. It begins when two young monks, Ngawang and Lodoe, hover over an older monk named Geshe

Yeshi, who sits on the floor. They begin to argue over the ordering of the planets. Tenzin translates snippets for me but it's hard to keep up. Geshe Yeshi smiles and waggles his head and fingers dismissively. The two younger monks persist. They take turns in their good-natured assault. Soon everyone in the room is on their feet.

The air fills with the percussive sound of hand slaps, the right hand moving forward after impact, fingers pointing at the protagonist for added emphasis. A body lunge accompanies the clap. It's like fencing with words instead of rapiers. They laugh as they do this. I'm fascinated by the social dynamics. Any geshe such as Yeshi is highly trained and senior within the Buddhist hierarchy, but deference is set aside and the debate plays out in the arena of ideas. It's amazing to see this level of intensity among men without any hint of antagonism.

More monks gather until there's a scrum of two dozen. The younger monks drape their arms around each other and there's jostling and mild shoving. Laughter comes in waves, and I'm sorry to be missing all the jokes. I can't see through the mass of bodies and I'm not tall enough to peer over the top. I clamber up Thupten's back but fall back onto the floor. We're at the nominal end of a half-hour activity and only one of the groups' timelines has been critiqued. My plan for the day has been derailed. For a teacher, this is nirvana, a locomotive of active learning. I'm happily superfluous, a grinning fool.

Our scale model of time from the first week of class made clear how recent humans are in the 4-billion-year pageant of life on Earth. Biological diversity stems from a single genetic code, but it was sent in myriad directions by the unguided hand of natural selection. The monks break into groups to explore one aspect of this. It's a low-tech experiment, usually done on computers in a laboratory, but we need only colored paper.

Each group gets a plastic sandwich bag filled with colored dots made by using a hole-punch on yellow, green, blue, pink, and purple sheets of paper. Those are their creatures. They also get a sheet of paper of one of the five colors. That's the environment. The rules are as follows. They put four dots of each color randomly on the sheet, a total of twenty. One monk acts as "predator." Having turned away as the dots were being placed, at a prearranged signal they try and take or "eat" ten dots as fast as they can. Then each of the remaining ten dots reproduces, so another dot of that color is added, making twenty dots. The predator does its cull again, eating as fast as possible, and the surviving dots reproduce one more. After five generations, the experiment stops.

As the groups complete the first generation, it seems too easy for the predators. They're dining with impunity. So I tell them they must use their nonwriting hands to eat. I also ask a second monk in each group to act as a distracter, making life harder for the predator, just like in the wild. That works well—the predator monks reach in to grab dots but often come up empty-handed and the distracter monks enjoy their role as spoilers, waving their hands in the faces of the predator and occasionally tickling them.

At the end of the experiment, we look at the outcomes. They're dramatic; 80 percent to 100 percent of the dots in each case are camouflaged—yellow dots on a yellow sheet, pink dots on a pink sheet, and so on. The rest have been culled by natural selection. Very few dots survive that don't match their environment. The message is clear: you can be colorful to attract a mate, but not if there are any large predators with shaved heads and red robes around.

What's so special about Earth, I ask the monks as we once again bring out our small metal sphere painted with blue oceans and brown land masses. I hold it up. Carl Sagan famously called the Earth a "pale blue dot" in a vast universe. How unique is this place we call home, with its diverse and persistent web of life? And how special are we, with our hopes and fears and aspirations? In answer, I think of the Roman poet and philosopher Lucretius. Two thousand years ago, he boldly speculated, "It is in the highest degree unlikely that this Earth and sky is the only one to have been created. . . . So we must realize that there are other worlds in other parts of the universe, with races of different men and different animals."

Lucretius and other thinkers dreamed about other worlds long before we had the technology to address the question. Finally, the answer is just around the corner. Theorists have long expected that the circumstances that formed the Earth are not unique. Planets should form as natural byproducts of star formation. But how many planets and what kind? Until recently, nobody knew, and the observations were fiendishly difficult, as our scale model from the first week made clear.

After dinner, we have an extra session to go over the monks' journals. Then we go outside to understand the difficulty of seeing planets around distant stars. The foothills of the Himalaya are impressively dark at night. We walk through the campus gate and down the road a hundred yards to a place where no lights are visible. The monks chatter as we walk. We gather at a turn in the road where there's a promontory, and we can see into the river valley. Along the riverbank we see flickering fires of the impoverished Indians who live under awnings and in tents made of rags.

Our goal is to create a new scale model with a battery-powered headlamp that I brought from home serving as the Sun. I've removed the reflector so it's a naked hundred-watt bulb. At this scale, the Sun is ten centimeters across, and that would make the Earth the size of a bead. I've brought a white bead with me and attached it to a thread. "This is the Earth," I tell the monks. Where would you place it? We've done a similar scale model before so they should have the idea, but like many people they underestimate the emptiness of the solar system. Most of them want to put the bead close, a few feet from the light bulb. It should actually be ten meters away.

Dawa holds the Sun while Jigme paces off the correct distance and holds the thread above his head so that the Earth hangs at eye level. How far away can you get from the bead and still see it by its reflected lamplight? The monks test this out, moving closer and further from the bead. To make it easier I've brought a piece of black cloth, which Tenzin holds to the side to block direct light from the bulb. Even when we all know where it is, the bead is extremely hard to see. Such a small object catches just a tiny glimmer of light from the bulb ten meters away. The consensus is, "We can see it no more than a meter away." Beyond that the bead disappears into the black night.

But we're still within the solar system. If you wanted to see a planet like the Earth at the distance of the nearest star, you'd have to detect the bead from ten miles away!

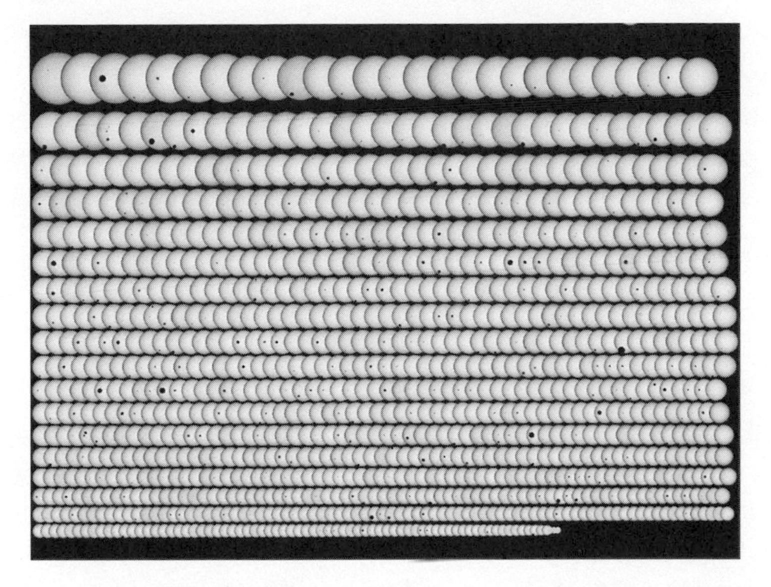

Figure 6.4. After dreaming of distant worlds for centuries, exoplanets were finally discovered in 1995. Recently, the Kepler satellite has identified thousands more by their subtle eclipses as they cross their parent star, as shown in this montage.

It sounds hopeless, and indeed astronomers looked for planets with large telescopes for decades without success. A planet's reflected light is too feeble to detect. Planets tend to be close to stars hundreds of millions of times brighter, swamping their faint reflected signals. Then, in 1995, astronomers detected Jupiter-mass planets around other stars by an indirect method—which I now demonstrate.

Behind us is a hill with a sparse pine forest. I keep Dawa and Jigme at the roadside with me and send the rest of the monks up the slope so they can look down on us. I fix the headlamp onto the top of Dawa's head with the Velcro strap under his chin. First I ask him to pirouette, like a human top, holding the thread with the Earth.

"Do you see anything?" I call up the hillside.

"Nothing," comes the reply.

Next, I hand Dawa a heavy rock and ask him to spin around with the rock cradled in his outstretched hands. I stand back to avoid getting hit in case the rock slips out of his grip.

"What do you see now?" I call to the monks out in the darkness.

"We see movement, motion. A circle. An oval shape," come the replies.

In the dark, all they can see is Dawa's headlamp tracing out a shape as he spins with the rock. Like a hammer thrower or a parent whirling with a small child, Dawa must lean out to balance the weight of the object he is spinning. Seen from directly overhead it would be a circle; seen at an angle by the monks on the hillside it's an ellipse. To exaggerate the effect I ask Dawa and Jigme to grab each other's wrists and spin around.

"Now?"

"A bigger shape," reply the shadowy figures.

The next day in the classroom we cement the lesson. I have a battery-powered buzzer on a strong string and I turn it on and swing it in circles above my head. It's a demo of the Doppler effect. When the source of waves moves toward you, it catches up with its waves, shortening the wavelength and increasing the frequency or pitch. When the source is moving away from you, it recedes from its waves, making for a longer wavelength and a lower frequency or pitch. The monks hear the whine of the buzzer oscillating between a higher and a lower pitch as it orbits above my head like an angry bumblebee. The Doppler effect applies to any wave, so it describes a moving source of light as well as sound.

Now we can understand the new method that astronomers discovered to find tiny planets orbiting other stars. We call these "exoplanets." They are detected only because of their influence on the parent star. When planets orbit stars, the star isn't immobile; they both orbit a common center of gravity. For Jupiter and the Sun, Jupiter tugs the Sun into a pirouette around its edge, taking twelve years, the period of Jupiter's orbit. Exoplanets were found the same way. As we saw the previous night, the unseen smaller mass tugs the visible large mass into a circular motion, and it's detected by looking for a periodic Doppler shift in the starlight. The bigger the planet, the bigger the effect.

After the initial hard-won success of this insight for astronomy, the floodgates opened. Precision spectroscopy—using an optical spectroscope to disperse and measure the spectrum of light—was used to dig out the subtle signals from weeks and months of observations of

Sun-like stars. The number of exoplanets passed one hundred in 2004, a thousand in 2010, and three thousand in 2013. Techniques have advanced to the point where we can detect planets that are as small as the Earth. The recent surge in exoplanet numbers was driven by a modest one-meter telescope launched by NASA in 2009 that took data until 2013. The Kepler satellite detected exoplanets by the fleeting shadow they cast on their star when they pass in front of it, an effect that only occurs if the plane of the orbit is along our line of sight to the star.

At the tea break I demonstrate this shadow effect in the hallway where there's a large globe light hanging from the ceiling. With a disk of black cardboard cut to be 10 percent of the diameter of the light fixture, I attach it to a long wooden rod and pass it in front of the light. "Can you see that?" I ask. Murmurs and nods of assent; they can see the effect, but they can't detect it since the disk only eclipses 1 percent of the light from the fixture. "That's easy to detect with a ground-based telescope," I say. With the stability and freedom from atmospheric effects of a space-based telescope, NASA's Kepler can detect eclipses of 0.01 percent. Since the Earth is ten times smaller than Jupiter, that's how you detect Earthlike planets orbiting distant stars. A small fly buzzes lazily in front of the light. I point to it, "That's like an Earth transit."

I summarize the results of this thrilling research. There are over three thousand exoplanet candidates. Several hundred of these are Earthlike, and of these, a few dozen are in the "habitable zones" of their parent stars—Goldilocks orbits where temperatures could allow water or biological life to survive. The data are consistent with all Sun-like stars having Earthlike planets. It projects to several hundred million habitable worlds in the Milky Way and a jaw-dropping billion billion (or 10^{18}) in the universe. Tenzin carefully translates these large numbers into Tibetan.

I pause to let this sink in. The room is completely silent.

Before we break into groups to discuss the implication of so many Petri dishes (that is, rocky exoplanets) where biological experiments may have occurred, I pass around a thimble of sand and a magnifying glass. This isn't the builder's sand we used in our estimation experiment, which was plain quartz. This is sand from my favorite shell beach

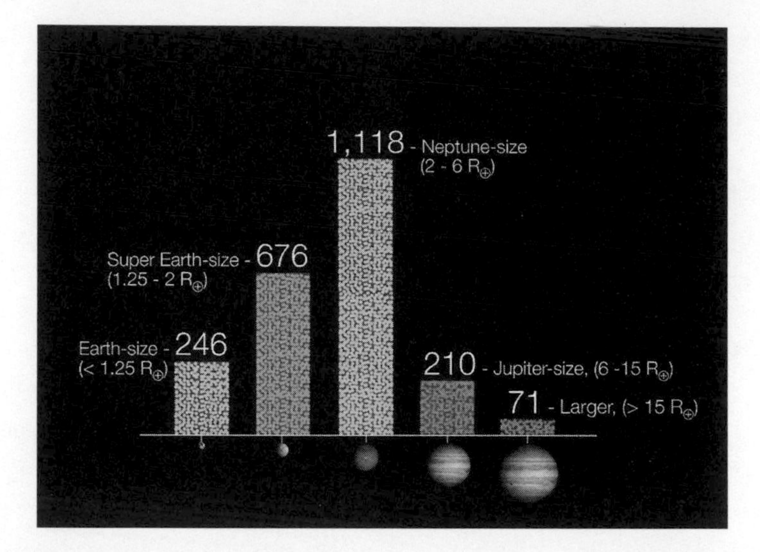

Figure 6.5. The Kepler satellite can detect exoplanets that are Earth-sized or smaller. By mid-2013, the census included hundreds that are terrestrial and more than fifty in the habitable zones of their parent stars.

in southern England. Under a magnifying glass the multicolored dots resolve into miniature worlds: cone-shaped, spiky, scalloped, whorled, beaded, soft like chalk, hard like obsidian, all the colors of the rainbow. This is how many worlds we've discovered so far, and we haven't inspected any this closely. The number of worlds in the universe is all the grains of sand from all the beaches of the world.

The Fifteenth Dalai Lama

Awesome! Fantastic! Amazing! At breakfast, Karma is gently mocking the tendency of Americans to use superlatives when they're not always warranted. Tibetans prefer to avoid emotional extremes. High praise might be "pretty good" or "not too bad."

Everyone's in good form. Bryce is the King of Coffee, dispensing quips and brown elixir. Gail is pleased with all the progress the monks have made in her biology class; Richard and Mark are busy evaluating the program. They've successfully instilled journaling skills in a group with little writing experience. The evaluators work well together, but in some ways they're an odd couple. Richard is from England and Mark is from California. Richard's crisp and military bearing riffs off Mark's laconic and laid-back vibe. I imagine them as they might have been twenty years ago: the Squaddie and the Stoner.

I'm wearing a T-shirt today with a crudely drawn cartoon of the Dalai Lama opening a box for his birthday. He says. "Wow! Nothing, just what I always wanted." Everyone takes a moment to read it. "You're lucky we are liberal-minded," says Lhakdor. "We should give you a good beating." He pauses a beat. Then he grins.

I feel a twinge of shame; to a Westerner it's no big deal but a Tibetan might find my T-shirt disrespectful. Lhakdor brings us down to Earth. "We give too little thought to the consequences of science," he says. "We're too careless with the planet. Politics panders and is mostly free of substance. In India, it is a government of the cattle, by the cattle, and for the cattle. The bad people will get the weapons." After this bleak prognosis, silence falls around the table.

"God, that's depressing," says Richard.

"On the contrary," says Lhakdor brightly. "I'm an optimist. It's the only way to be."

To prove the point, he tells a joke. A high lama visits a remote place for a tantric celebration. Many people come to be blessed and he can't bless them all, so he blesses a few in the front row and tells the rest of the crowd to imagine they've been blessed. One man is angry. He's brought a gold coin as a gift for the lama but he's not one of the few blessed. He says to the lama, if I have to imagine I've been blessed then you can imagine I gave you this coin. Another man is behaving very badly by drinking alcohol in front of the lama. The lama says, haven't you been taught to keep your mind free of intoxicants? The man replies, I'm pretending that the wine is water and that I'm a donkey. What's wrong with a donkey drinking water?

Almost everything Lhakdor says is a joke on the surface, but with a deeper meaning or insight contained within. It's a disarming strategy. He gets up, strolls around the table, and puts Karma in a head lock, mocking him for his slight frame and modest musculature. Then he leaves the room smiling and flexing his biceps.

It's the penultimate day of the workshop. Euphoria, exhaustion, and sadness mingle inside me. I will miss this group of extraordinary students of the universe and the purity with which they approach learning. I've been handing out NASA stickers liberally. On the last day, I dispense something special: rub-on tattoos of M51, the Whirlpool Galaxy. Most apply them to their biceps or forearms. A few get bold and emblazon their forehead or cheek. It takes a while for everyone to settle. When they do, I pose one final provocative question.

"What if you woke to find you were the fifteenth Dalai Lama?"

The Dalai Lama is believed to be the rebirth of a long line of spiritual masters who are considered to be manifestations of the Bodhisattva of Compassion. This lineage of the Gelug school of Buddhism started in the fourteenth century, and since the third incarnation, the Dalai Lama has had Gyatso or "ocean" as a part of his name. So far, the Dalai Lama has always been a man and always born in Tibet, but the current Dalai Lama has not ruled out that the next one might be found outside Tibet and might be a woman.

By asking them to imagine being the chosen one, I hope the monks will accept the hypothetical situation as a purely intellectual exercise. Suppose you were a young boy, and one day you learned or were told that you were the next Dalai Lama. How would you feel? Would you be amazed at your good fortune and intimidated by the responsibilities of being such a major religious figure? Or would you say to yourself, *Well, among 6 million Tibetans, somebody has to be the Dalai Lama, so why not me?*

By this conceit, I'm able to introduce the ideas of "fine-tuning" and the "anthropic principle." Is our planet with sentient humans—*anthropos*—so remarkable that it can't be a fluke? Or is it that some planet among billions was bound to have life, like a throw of the dice, so it may as well have been Earth? By fine-tuning, astronomers mean that all the strengths of forces and values of fundamental constants have led to a universe with carbon-based life when, if they had been slightly different, our world would have been an inert and uninteresting cosmic soup. The problem for physics is why we have these forces of nature and these constants and not others. The enigma deepens when we realize that the constants produced human beings who are asking, "Why do the constants have the values they do?"

This is a why question about the universe, so it may sound metaphysical, but it's not like asking the question, "Why is there something rather than nothing?" Fine-tuning is simply concerned with various physical relationships. Consider a few examples. There's a specific ratio of the strength of electromagnetism to the strength of gravity (a huge number), a specific ratio of expansion energy to gravitational energy in the universe (close to 1), and a specific value of the energy density of empty space (a tiny number). Another example is the very slight asymmetry between matter and antimatter that led to all the stuff in the universe. Or consider the fact that space has three dimensions.

All these circumstances are presumably explained by a "theory of everything" that's not yet within our grasp. Meanwhile, if these values and ratios were different, our universe would be unrecognizable. Indeed, there wouldn't be anybody there to recognize it. For example, if contraction energy from dark matter had exceeded expansion energy from dark energy, the universe would have lasted too short a time to form stars and life before it collapsed. And if the expansion energy had dominated, gravity would not have made stars, and the universe would have ended up as a cold and diffuse gas. If physical constants and forces of nature had somewhat different values, there would be no stable atoms, or no molecules, or no long-lived stars. These "hypothetical universes" have laws of physics and so aren't random or nonsensical; they're just universes without biology, without *us*.

The great debate over the anthropic principle—which suggests the universe is somehow built for life—was triggered by a paper by the English astronomer Brandon Carter on the occasion of the five hundredth anniversary of the birth of Copernicus. That was a suitable birthday to raise this topic, since the anthropic principle seems to reverse the "Copernicus principle," which consigns us to a position of mediocrity in the universe. The anthropic principle asserts that we are in fact special, and not just a throw of the dice or a cosmic accident.

One response to the anthropic principle has been the idea of a "multiverse," because if there are multiple (even infinite) parallel universes,

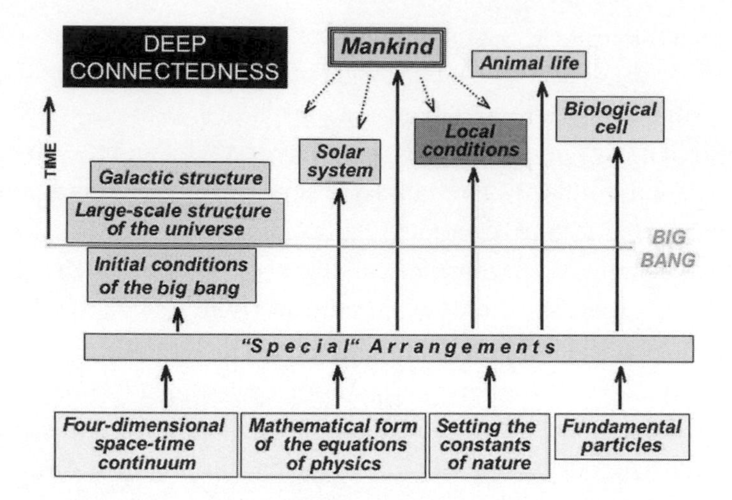

Figure 6.6. There are aspects of nature and physical constants that are propitious for the formation of carbon, long-lived stars, and biology. In a multiverse scenario, most of the universes would probably be devoid of life.

then a cosmic throw of dice could certainly have produced a world like ours. Our existence is then not surprising or special; it's simply a matter of probability. Using the multiverse theory to defeat any anthropic belief in Earth's special origins has its own problems. The biggest is that we can't observe other universes, so we can't investigate the idea. We can only observe a universe with properties that permit our kind of biological life to form.

Opponents of the anthropic principle say the idea is sterile. It can't predict anything that science can test. They also point out that fine-tuning makes assumptions about what's needed to produce life and sentience. What if life or intelligence doesn't need carbon or water or stars? English writer Douglas Adams poked fun at this way of thinking with the parable of a puddle who is amazed by the coincidence that each time it rains, the hole fits him perfectly! Similarly, foolish humans are amazed that they're on a planet where the food and air perfectly suits our needs.

Now I ask the monks again, how do you react to the news that you are the Dalai Lama? Most accept the outcome as serendipity. Geshe Nyima would be honored and feel incredibly lucky. Dawa says he'd be shocked. Jigme adds that he'd have trouble accepting the information.

So it is when we consider our role as sentient observers of a vast and inhospitable universe that seems to nevertheless fit us perfectly. We can appeal to providence: being the Dalai Lama or living in a universe made for us is the result of a divine plan. Or we can say it was necessity: *somebody* has to be the Dalai Lama and we can *only* live in a universe that's hospitable to life. Or we can say it's chance: one in 6 million Tibetans is the Dalai Lama, and our universe is the rare life-bearing one within a mostly sterile multiverse.

Yet none of these arguments capture the special feeling we have as humans. The Buddha once told his disciples, "Imagine the whole Earth was covered in water and that someone threw a wooden yoke with a hole in it on to the surface. And suppose that once in a hundred years, a blind turtle were to rise to the surface. What are the chances that it would put its head through the hole in the yoke?"

His disciples replied, "It's very unlikely, Lord!"

"Well," said the Buddha, "it's equally unlikely to be born as a human being." The Buddha was saying that being born human is an extraordinarily rare privilege, and one to be treasured.

At noon the next day we're all on top of a nearby hill, praying for the health and long life of the current Dalai Lama. It's Sunday, with no classes, and after breakfast we hike past twittering birds and chattering monkeys. The monks sing and chant and laugh all the way up, their burgundy robes playing perfectly against the deep green of the pine forest. At the top, the prayer flags are blessed before they're draped between the branches of two tall trees, joining tattered and faded remnants of previous ceremonies. Then we head to a glade and Lhakdor leads a heartfelt prayer for His Holiness. As I listen, the wind carries a scrap of an old prayer flag toward me and it flutters in front of my face. I reach out, grasp it, and put it in my pocket as a talisman.

While the ceremony continues, my mind wanders to the only time I've seen Tenzin Gyatso, the fourteenth Dalai Lama. It was four years before I got the call to join the Science for Monks program, when he spoke at our Tucson Convention Center. I recall one of the questions he was asked from the audience. His Holiness had been talking about

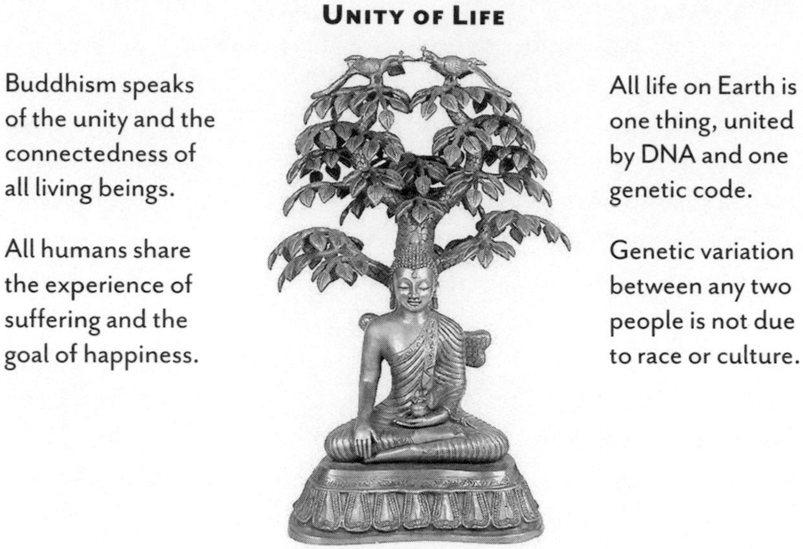

Unity of Life

Buddhism speaks of the unity and the connectedness of all living beings.

All humans share the experience of suffering and the goal of happiness.

All life on Earth is one thing, united by DNA and one genetic code.

Genetic variation between any two people is not due to race or culture.

Figure 6.7. Modern biology speaks to the unity of all living things; from bacteria to elephants, they all share the same genetic code. The genetic variations among different human races are smaller than the variation within a race.

mindfulness and meditation. Then a man stood and talked about how he had studied Buddhism for years and had seen benefits from meditation. But he was a busy executive, and with all of his commitments he couldn't find the time to meditate for more than an hour a day.

The question was translated, and the Dalai Lama replied in English in a light, lilting voice, "I'm quite busy, too."

At our ceremony on the Indian hillside, the prayers and chants have concluded. Some monks are showing off their M51 tattoos, which are already fading. I have a permanent tattoo on my left bicep, and it may take longer, but it too will turn to dust. To my surprise I'm beginning to be okay with impermanence.

After the ceremony we have juice boxes and masala chips. Everyone sits in a large circle on the grass and we play a children's game. One monk runs around the outside of the circle and places an empty juice box behind one person's back. They can't turn around; they can only feel for it with their hands. If they find it, they have to chase the monk who put it there, catching him before he reaches his spot again. If he's caught, he must pay a penalty. If not, the box is put behind someone else's back. Some of the larger monks move with amazing speed and agility. Jigme and Dawa race round the circle like dervishes; nobody can catch them. Penalties include singing a song, doing a dance, and acting like an animal. Nyima is caught and dances with abandon in the center of the circle. Bryce croons "Amazing Grace" lovingly to Karma.

Abruptly, Lhakdor claps his hands and prepares us to leave. "Too much of any one thing is not good," he says.

At One with the Universe

The three weeks have gone by in a blur, and I'm wondering how the monks could possibly absorb all the information we've covered. I also worry that subtleties have been lost in translation as cosmology bridges two cultures and two languages. It's ironic that teaching is more challenging at home where the culture and language are shared. These Tibetan students have walked with me each step of the way, their enthusiasm and curiosity undiminished.

For the last evening, the night before graduation, we'll do something different. The monsoon has abated, and the last few nights the sky has been ablaze with Byzantine waste. We've spent too much time sitting in the classroom. As Walt Whitman wrote, "When I heard the learn'd astronomer; when the proofs, the figures, were ranged in columns before me . . . how soon, unaccountable, I became tired and sick; till rising and gliding out, I wander'd off by myself, in the mystical moist night air, and from time to time, look'd up in perfect silence at the stars." Tonight we'll tilt our heads upward to the sky.

In the last morning session before lunch, I bring seven rectangular boxes in on a cart, one for each monk group. With the eagerness of kids on Christmas morning, the monks tear open the wrapping paper and unpack seven Galileoscopes.

The Galileoscope was released in time for the four hundredth anniversary of Galileo's first use of the telescope. It updates his best design with modern components. The lenses, tube, and mount are made of hard, well-machined plastic. The cost is just twenty-five dollars. Each group also gets a tripod, since the telescope has a small field of view and it's too long to hold steady by hand. The groups work swiftly and test their assembled telescopes by pointing them across the courtyard. Several groups have been too impatient to read instructions properly so they're unable to focus on a distant object. The more successful groups laugh, then go over to help. On their tripods, the telescopes are set at the side of the room to wait for nightfall.

At this point we've reached a certain limit in explanations. We've done the science. Later we will simply look out on the universe. By falling silent, we can all think about what's left—the meaning we want to give to what we see. I'm not sure if science can help with meaning. Science is still in its childhood, an unruly teenager of great promise that doesn't know its own strength. It's capable of great good and it also has the potential for harm. As a singularly human enterprise, science reflects all our strengths and faults.

In class, we return to our original theme of the limits of knowledge. In our first few days together we saw that there are no perfect observations. We're blocked from perfect apprehension of the microworld due to quantum uncertainty. There are also severe limits to our knowledge of the macroworld. Complexity is a pervasive feature of physical and biological systems. All this ambiguity means that determinism has no place in modern science. Nor does reductionism. In these acknowledgments, science is in harmony with the Buddhist worldview. Rather than being the predictable sum of its parts, nature is a source of surprise and delight.

I pose a question to the monks: "Water is clear, wet, and a solvent. A single water molecule is not. At what point does water become water?" Then another: "Bacteria can adapt, reproduce, and evolve. Their chemical constituents cannot. At what point does something become alive?" And a third: "Brains can think, be self-aware, create, and know death. The neurons that form a brain cannot. When does matter become sentient?"

In each case, the higher level of organization has emergent properties that are not predictable from the behavior of the constituent

Emergence is the idea that the whole is more than the sum of the parts, just are we humans are more than the chemicals we're made of.

Figure 6.8. Strict reductionism fails in science because properties often emerge at higher levels of complexity and organization that aren't apparent at the lower levels. This idea is called **emergence**.

parts. If enough water molecules are brought together, they do indeed become clear and wet, dissolve other chemicals, and form the complex shapes of snowflakes when frozen. That example is not mysterious. But scientists can't simulate the transition from a complex, concentrated chemical network to an entity that partakes in Darwinian evolution, and even the best laboratory in the world can't take a complex set of biochemical or electrical switches and make them think. Are the processes that turn brute matter into living organisms and mental life ultimately mysterious? Scientists don't think so. Yet there's no theory to predict when and how those transitions occur.

"Do bacteria feel joy?"

This question comes out of the blue from Thupten. Like many of his questions, it's surprising and provocative. I use it as a bridge to an activity where we put the names of species on note cards, ranging from elephants and dolphins to rats, birds, moths, worms, and bacteria. I ask the monks to work in groups and lay the cards out in order of intelligence; I deliberately don't define *intelligence*. Then I ask them to divide the species into sentient and nonsentient. The monks draw different conclusions—some say dogs and cats are sentient but not trees. Others include plants. *Sentience* is defined in the West as the ability to be conscious and have subjective feelings, but in Tibetan Buddhism sentience is often assigned to "lower forms of life" or even inanimate objects.

The standard scientific definition of life is too doctrinaire. Biologists would tell Thupten, of course not. Bacteria "feel" nothing. They have no sensory apparatus, no central nervous system, and no brain. They're too primitive to be sentient. On the other hand, bacteria do show surprising behavior. They use *quorum sensing* to orchestrate chemical communication in a colony, and their relationships with other organisms are complex and symbiotic.

It seems to me that science really can't define the necessary or sufficient conditions for consciousness in terrestrial biology, let alone for biology of alien functions and forms. Our ignorance should make us humble. There's no reason that the qualia that mark human experience—the sound of a flute, the color of a sunset, our sense of disgust, or the sharp recollection of a lucid dream—should be unique to denizens of the Earth. Life has no doubt evolved on a fraction of the habitable worlds in the cosmos. We're not necessarily the first or the most advanced life form, especially if we consider that the universe had been evolving for 9 billion years before the Earth formed.

Some scientists have speculated that life and intelligence might take purely computational forms, not dependent on the carbon chemistry of our world. After all, our current technological path may allow us to make artificial and computer-based life forms. A few philosophers have even argued that human beings could be the simulated, computational playthings of a hyper-advanced civilization. Our rocket travel into space is actually quite primitive when compared to our own digital technology, in which information—and therefore a coded form of life—can transmit from A to B in seconds. Making life that's purely digital would be a natural evolutionary stage. If an alien civilization had created computational entities like us, the nature of computation makes it easy to produce vast numbers of them. Computational life forms would very quickly outnumber biological life forms. Based on a Copernican argument, we are unlikely to be one of the small minority of biological species.

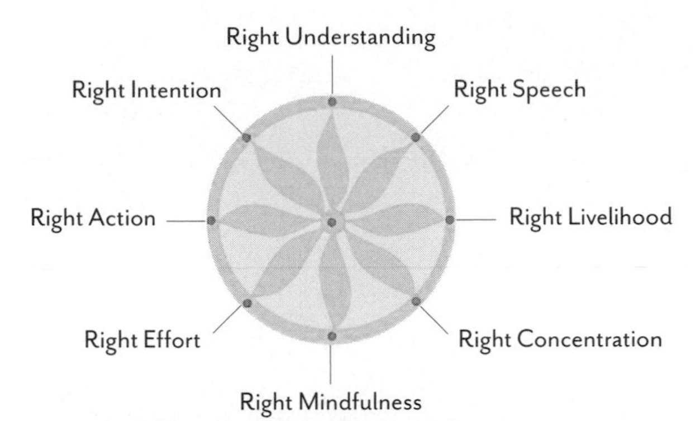

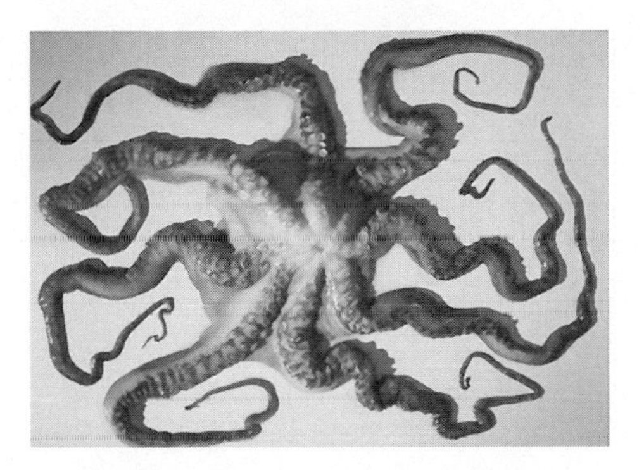

In the universe with 10,000 billion billion stars and probably a myriad of life forms, we are special in some ways, yet we are not in a cosmic sense. The profound question science is unable to answer:

Why are we here?

Figure 6.9. Science answers the "how" of life and the universe but not the "why." It's unlikely that we're the only or the most advanced life forms in the universe. Discovery of life beyond Earth will impact our self-image.

I see incredulity on the monk's faces. Surely I'm not serious. Maybe not. But when I ask the monks to debate the idea in their groups, they find it's a very difficult hypothesis to disprove. Geshe Nyima says that in his tradition there's no bound placed on the number of living worlds. They're infinite in number and are eternally going in and out of existence. Furthermore, there's no insistence that humans are the most advanced life form. Buddhist cosmology has the idea of a trichiliocosm, a system of a billion worlds, and a highly enlightened being, or supreme *nirmankaya*, with purview over that number of worlds. It occurred to me when I first heard this that the concept maps perfectly to the likely number of habitable worlds in a galaxy and the most advanced life form in that galaxy. I doubt that it's us.

How do I pull all these strands together?

It's impossible, so I end the day, and the official classroom sessions of the workshop, by telling them about the dream I had the night before. In my dream I was back at the big bang. The moment was filled with expectation; the universe has taken a breath and is poised to speak. I look out at the smiles and nut-brown faces and know why I had that dream. My particles and the particles of each of the monks have taken an amazing journey through the universe, fleeing from the coruscating heat of the big bang into the shock of night, buffeted by nebulae and floating in the interstellar void, riding the Milky Way's carousel, passing through the fusion cores and atmospheres of numerous stars, and then churning through an active geological world before coming together into my body and the bodies in front of me. From the unity of the big bang to the unity of biology. We are multitudes. Our atoms diverged from a single point and have taken myriad paths through space and time before arriving here to reflect on this moment.

We're all interconnected, and at one with the universe.

○○○○○

It's raining five or six hours a day. The relentless monsoon is powering a plague of insects. Despite Paul securing our perimeter defenses, we have to get rid of half a dozen cockroaches and an assortment of flying and biting bugs before we go to bed.

With the lights out, we continue our freewheeling discussions about astronomy. He's intrigued by the logic that we're unlikely to be the first or the most advanced technological civilization in the galaxy. So it's entirely possible that we're simulated entities. We talk about how best to behave if you think you live in a simulation. There seem to be three choices: act like a nihilist because nothing matters, follow the Golden Rule, or act interesting in case they decide to write you out of version 2.0.

Paul has a friend who's schizophrenic but is not taking his medications. The friend talks about the "system" as the pervasive construct, and even if it's mostly the meds talking, his assessment still sounds about right to me. The people we call crazy may be the only ones who see to a deeper reality—just as they occasionally exercise a full (and frightening) level of creativity.

I wake up in the middle of the night, and there's a green flashing light projecting on the ceiling. It's not from a clock or phone. When I find the source it's so weird that I wake Paul up. There's the carapace of a large beetle under the bedside table. Even though it has no legs and no head, the tip of its abdomen is emitting a bright pulsing light. I know about photoluminescence but this beetle has been dead for hours. I speculate that it's a communication back to the mother ship. We both laugh nervously. Finally, we drift off. When I wake, Paul has his back to me. He rolls over and I gasp to see his bristly facial hair and huge segmented eyes. Then I really wake up.

After three weeks in India, my stress has almost completely melted away. I haven't drunk alcohol for three weeks and I'm a temporary vegetarian. The uniformity of the diet and the rhythm of waking at sunrise and going to bed not long after sunset are a balm. I've enjoyed my time with Paul. We have enjoyed our mode of deciding in the moment what we want to do. Everything is fresh. I'm loosely tethered to my mind and my body. When I go home I suspect I will look at my life through new eyes. Assumptions will be questioned. The box will be opened and limits will be tested. Some of my mandalas will be swept away. Others will remain, but I'll no longer be as attached to them as I was before I came here.

Meanwhile, I know what I want to do.

Paul prefers to hang out in Dharamsala. So I look for a coconspirator, and Tenzin is game. He has business to do at the Tibetan Children's Village in Suja, and it takes little persuasion to get him to add on an adventure—paragliding on the edge of the Himalayas.

After the two-hour drive, we make a booking in the little town of Bir. With an hour to kill we get haircuts for a dollar each in a corner barber shop that's literally a hole in the wall. After the trim, my barber pummels my head with the heels of his hands; he's throwing in a head massage for free.

At the paragliding base camp, Tenzin and I meet two Indians. They look like teenagers; I wish they were older and more experienced. They drive us for an hour through pine forests to a high, wind-swept ridge. They lay out the chutes and the harnesses. These two youngsters will be our pilots, harnessed with us into the great glider sails. They bark out the instructions. "Run forward, and even if the harness pulls back, keep running!" Soon I see why. The wind rushing up the steeply sloped meadow creates such an updraft that the paraglider is carried upward and backward. Tenzin soon becomes a receding dot.

My turn. I'm in a harness in front of the pilot. He wrestles with the twisting sail and we scramble down the hill until my feet are treading air. The updraft hits us with a sickening tug and we're carried upward and backward. I look between my legs a minute later and am amazed to see the launch site hundreds of feet below. We've been flung into the sky. I now understand why the drunken Russians never returned.

Between my dangling feet, the Kangra Valley is a patchwork quilt of villages and terraces and temples. We've gained so much height I can see far into the snow-covered peaks. The harness and ropes creak and groan like a ship's rigging, and the sail is sensitive to the slightest tug of the pilot's hand. Then he yells "High G!" and tugs us into a steeply banked spiral turn that makes my sinuses stream. My mind drops to the pit of my stomach. Below us the terraces spin so fast they're a blur. Air screams past the sail, and I worry that the stresses will rip us from our harnesses and send us plunging to the ground. He repeats this maneuver several times, and I realize the thermals are so strong in the summer that this is the only way to lose altitude.

Minutes later we glide in to land on a fallow terrace. Almost instantly the clarity of the experience starts fading. But at least for a while, I was suspended a mile above the Earth, open to the sky, buffeted by the wind, and face to face with the vastness of the Himalaya.

How to end? Left to the Tibetans there would be no ending. We'd all disperse and head off to the next thing. But Lhakdor has pity on the Western teachers. He takes us out for a nice dinner, and he hosts a ceremony where Gail and I hand the monks certificates we've made. Gifts are exchanged. We bow as he drapes us with silk scarves or khatas—the white color symbolizes the pure heart of the giver.

He gives me his card. It just says "Lhakdor." He only has one name and in my mind it's because he's a kind of rock star, like Sting or Madonna. The card has his e-mail address, but with a serious face he warns me to keep any messages very simple because, he says, "Monks don't do attachments." He leaves in his inimitable way, pausing for one more story. A senior Tibetan lama visits New York for the first time and wants to eat like the locals. As soon as the airport taxi arrives at the hotel in Manhattan, he looks for the nearest hot dog vendor and says, "Make me one with everything."

Paul and I prepare to leave. The monks stop by as I'm packing up my teaching supplies. They say good-bye and we hug and exchange contact information. The oil of science and the vinegar of Buddhism don't mix completely, but they have been whisked together harmoniously. It's sad to leave new friends and head back into the maelstrom of work, and parts of my life that may taste more like vinegar than wine. But I recall a story about the great Eastern philosophers. They were tasting vinegar, which symbolizes the spirit of life. Confucius is the first to drink. "It is sour," he says. Laozi goes next and he pronounces the vinegar "bitter." Last to taste is the Buddha. He exclaims, "It is fresh!"

The Seven Ages of Monk

When I think of teaching the monks, my mental image is of a bubble growing, airy and buoyant, scattering light as it rises. I recall the time we used balloons as analogies for the universe. As we measured the expansion rate of the curved surface, some of the balloons burst, a first and second, then a third. Suddenly the room was a kaleidoscope of color as the monks batted around so many more balloons above their heads. It was a metaphor for their lightness of being.

That lightness is compromised only by their lack of a country. Tibetans face suppression of their religion and negation of their culture, and they've accepted that burden with fortitude and grace.

Buddhist monks aren't monolithic; they embody multitudes. All have taken vows that guide their behavior and mental habits, but they're allowed to follow different paths within the tradition. Some focus on traditional religious texts, some are politically active, and some venture beyond monasteries to serve local communities. The group I worked with all have a thirst for knowledge and are convinced that the conversation between Western science and Eastern religion will benefit both sides.

These Tibetan monks are unfamiliar with Shakespeare, but I see a connection to the monks in one of his most famous works. In the play *As You Like It*, the speaker opens with, "All the world's a stage." After that, the monologue goes on to survey the seven stages of a man's life: helpless infancy, schooling, unrequited love, soldiering, the wisdom of judgment, the shrinkage of old age, and finally the helplessness of late dotage. During my time in India, Shakespeare's evocative description came to mind as I watched the monks learn. We approached science as a form of play and even staged scenes from the universe. Along the way, I witnessed the "seven ages of monk."

There was Dawa Dorjee, naughty and mischievous as a child, his boisterous energy driving the class. There was Jigme Gyatso, with his satchel and shining morning face, creeping into the classroom. The memories of many more monks return easily enough. I have an image of Gelek Gyaltsen, strenuous in debate and sighing like a furnace. Then there is Tashi Phuntsok with his sharp intelligence and flashes of ego. I think of Geshi Yeshi in his fair round belly, full of wise saws and modern instances, and Geshe Nyima, with spectacles on nose and pouch on side, his big manly voice turning once again toward a childish treble.

No monk whom I had met had reached in mind or body the last age of man—the feebleness of old age. But even when they do, the Buddhist philosophy about the wheel of life assures them that they will not meet oblivion, but instead rebirth and continuation.

My time in India proved to me again that teaching is a wonderful job, and that teachers, especially at the college and university level, have many benefits and privileges. Nevertheless, education needs constant vigilance to prosper in our modern culture. The litany of problems in Western education is well known, especially in schools that prepare our children for college and the world: Teachers are poorly paid and disrespected. Schools are hamstrung by insufficient funding. Science teachers aren't always trained in the discipline they teach. Colleges aren't training enough scientists and engineers for the jobs that await them in a high-tech economy. In the United States, the public is supportive of science but has a low level of science literacy. Politics has marginalized the scientific consensus on issues such as evolution and climate change.

When we hit a trough, some Western educators look to the East and ask how education works there and what lessons that might bring. Some educators lament that Western kids get the message that success comes only to those who are born smart, as if learning is an innate ability. In the East learning is about effort. Struggle and suffering are seen as routes to success rather than indicators of weakness and failure. One comparison of American and Japanese students by a psychology professor at UCLA suggests the consequences. First-graders in America and Japan were given an impossible math problem to solve. American kids worked on the problem for thirty seconds on average and then gave up. The Japanese kids worked for an hour on the impossible problem.

Science for Monks is a modest experiment taking place on another continent and bridging a gulf of language and culture. What could it possibly have to say about education in the United States? Plenty.

For educators, the message is that learning is a labor of love and joy. Students need to be actively engaged and encouraged to roam outside the box. In the United States, where problems in the educational system seem intractable, the solutions are actually simple: honor teachers, train them in their discipline, pay them properly, give them the resources to succeed, and hold them and their students to high standards of commitment and application.

My experience teaching Buddhist monks was inspiring because they were completely invested in learning. Buddhism has four premises, or "noble truths." The first is the fact of human suffering. The second asserts that a major cause of suffering is ignorance. Education is an antidote for ignorance and the monks and I shared that commitment. But Buddhism is a salutary worldview because it's not purely passive or contemplative. The story is told that a man came to the Buddha in great distress because he had been pierced by an arrow. He was very agitated and bleeding badly. He peppered the Buddha with questions about his predicament. But the Buddha cut him off, saying "Why do you ask such questions? Is it really important to know who fired the arrow, what kind of bow was used, or how fast the arrow traveled? You are bleeding; would you rather get answers to all your questions or seek medical treatment?" Knowledge is important, but it's useless without action.

For scientists, this modest program offers a lesson about reconnecting with the childlike wonder and curiosity that are the wellsprings of science. And in a world where science can lead to harm as well as benefit, we need to ask not only whether something *can* be done but also whether it *should* be done. Even as we seek to understand the universe, we need to be humble before the void. I'm reminded of the closing words of the Dalai Lama's Nobel Prize lecture. In India, I've been privileged to meet people who live these words every day:

> For as long as space endures,
> And for as long as living beings remain,
> Until then may I, too, abide
> To dispel the misery of the world.

Afterword

WHEN THIS BOOK APPEARS, it will have been six years since I spent my first three magical weeks near the roof of the world. The group of monks from that opening session in India graduated after their fifth workshop, in May 2010. I was there again to teach. We had an emotional closing ceremony presided over by the Tibetan Minister of Education. A second cohort of twenty-eight monks, with the happy addition of four nuns, started the Science for Monks program in 2011, and I went to northern and Southern India to teach at their third and fourth workshops in 2013.

Alongside their science classes, the monks were trained in designing and constructing science exhibits by staff members from the Exploratorium in San Francisco and the Smithsonian Institution in Washington, DC. Using their new skills, the monks created an exhibit called "World of Your Senses." It debuted in Dharamsala then traveled to the Exploratorium in May 2012. Several of the geshes in the group have started science centers in their home monasteries.

Geshe Lhakdor is the maestro of the program in India, and I thank him and his excellent staff for their support. Particular thanks are due to the translators: Tenzin Sonam, Karma Thupten, Tenzin Paldon, and Nyima Gyaltsen. Extra acknowledgment goes to Tenzin for conducting and transcribing my interviews with the monks, and for accompanying me on an aerial adventure.

Some of the people we met in these pages have passed through interesting transitions. Nyima Gyaltsen joined the program as translator.

Karma and Paldon got married and they have a baby. Tenzin and his family came to Tucson, and he's a PhD candidate in education at my university. The young man Ugyen, who still dreams of being the first Tibetan astronaut, went to England to study for an International Baccalaureate diploma, and he plans to apply to MIT to study aerospace engineering. Science has recently become part of the curriculum at all major Buddhist monasteries in India, and as might have been anticipated, almost all the monks have cell phones and most are on Facebook.

Lest we forget, however, the situation in Tibet is still dire. Since my first visit, at least a hundred monks and laypeople have self-immolated as a protest.

I'm grateful to friends and colleagues for their support over the years of this adventure. Bryce Johnson changed my life by inviting me to join the Science for Monks program, and he more than anyone has given it a special stamp of enthusiasm and commitment. Since then, Bryce has been employed by the Exploratorium, and he continues to organize Science for Monks, along with occasional symposia on science and religion, and a new program to give an overview of modern science to geshes from all over India. Another colleague who made this work easier was Ed Prather. Ed is a pioneer of learner-centered methods of instruction like those described in these pages. He preceded me in teaching in the program, and his enthusiastic report on the experience made me certain I wanted to go.

My younger son Paul was the perfect traveling companion for my first trip, and I'm sure we'll never forget that time together. My divorce, and the loss we both felt, has been healed by time. After the second workshop I met a wonderful woman named Dinah Jasensky, who not only warmed my heart, but also encouraged my writing endeavors. For his part, Paul graduated from my university in 2013. He's currently in Finland for graduate school and wants to be a teacher. (I've not forgotten that I owe my older son, Ben, an exotic trip.) I have other travel companions to thank, including my cousin Louise, who joined me on my second trip. The rest are my fellow teachers at workshops in northern India, where we shared camaraderie and insights: Gail Burd,

Richard Sterling, Mark St. John, Stephanie Norby, Lori Lambertson, Donald Gallehr, and Scott Schmitt.

I would like to thank the Sagar Family Foundation for steadfast support of the workshops and the Templeton Foundation for generous support of this writing project. My agent, Anna Ghosh, has been helpful in many aspects of my writing career. I'm grateful to Susan Arellano, editor in chief at the Templeton Press, for her enthusiastic support of this project, and to Larry Witham for his work on the manuscript. They both encouraged me to leave the familiar lake of popular science writing and venture into the wilder waters of narrative nonfiction. Larry in particular exercised a powerful form of alchemy on the first draft, helping me to alloy science and pedagogy with personal experience. Part of this book was written at the Aspen Center for Physics. I've made a mental note to find a way to bring the monks there; the blending of high-octane physics, thoughtful Buddhism, and natural beauty would surely be intoxicating.

However, the biggest thanks go to my maroon-clad students, whose joyful spirits motivate and animate this book.

Index